More Than An Engineer

By

Rob Ransone

Rob Ransone
PO Box 425
Wicomico Church, VA 22579
(804) 580-5929
ransone@aol.com

Some of the names have been changed
Either to protect the culpable
or because I can't
remember them

ISBN - 13: 978-1490481623
ISBN - 10: 1490481621

More Than An Engineer

I reach my dream job–Edwards AFB

I had graduated from Texas A&M in January, 1956, with a Bachelor of Science degree in aeronautical engineering and a U.S Air Force ROTC commission as a second lieutenant. I had finally received my orders to active duty at the Air Force Flight Test Center (AFFTC), Edwards AFB, California, about 65 miles north of Los Angeles. On July 3, 1957, I loaded my personal belonging and new Air Force uniforms into my 1956 Buick, and headed west. I drove from Fort Worth, Texas, to Amarillo, in West Texas, from where I followed U.S. Route 66 through New Mexico, northern Arizona, and into California. There were few Interstate highways then, and most of the highways were only two lanes. There were deep dips in the New Mexico and Arizona highways where the Route 66 crossed dry riverbeds. At least most of the time they were dry—occasionally, due to sudden severe thunderstorms in the mountains, walls of water would gush down the riverbed and flood the road without warning. Signs at the crossings warned of this, but there were still cases of cars being swept away and their occupants drowned. I bought a canvas water bag and hung it over the radio aerial on my front fender. Just enough water seeped through the canvas to wet the outside, and the dry wind blowing over it evaporated this dampness and cooled the water very nicely, thank you. In those days air conditioning was an aftermarket add-on and my car did not have an air conditioner.

Since I was on Air Force orders, I spent a free night on the road in the Bachelor Officers Quarters (BOQ) at Kirkland AFB, in Albuquerque, New Mexico. I was used to the parched, flat, barren countryside of west Texas and it didn't change much as I drove into New Mexico, so the long climb up over the mountains and then down into Albuquerque was an interesting change. Albuquerque is located at a much higher altitude than Fort Worth, where I had bought my car. I noticed this difference in altitude (Albuquerque is at about 7,500 feet, compared to Fort Worth's 600 feet altitude) when I tried to pass another car. Normally my car, a Buick Century with 245 horsepower, usually had no trouble leaving most cars behind, so the first thing I did the next day was to find a Buick dealership to see what was wrong.

The problem was with the idle jets in the carburetor—this was before most cars had fuel injection. The mechanics replaced the low altitude idle jets with high altitude jets, and I was quickly on my way, with power restored.

The trip through northern Arizona was delightful, up through the Rocky Mountains with the highway bordered by dense pine forests. It was cool up there, which was fortunate without air conditioning, but as I approached the Columbia River that divided Arizona from California, the road dropped considerably, only to begin the long climb up into the High Desert at Needles, California, once I crossed the river.

Now my car started overheating. The temperature gauge was approaching the red line, and I still had several miles to go before I reached the summit. What to do? I wasn't really worried because at worst I could pull over to the side of the road, stop, and then continue on after the car cooled down.

Relying on my engineering knowledge, I concluded that I needed to extract more heat from the coolant. The heater draws heat from the coolant water, so I turned on the interior heater full blast, and hung my head out of the window! It was blisteringly hot inside, but using the heater extracted just enough heat from the coolant that I made it to the summit without overheating the car, and the coolant temperature dropped back into the normal range once I reached the summit and the road leveled out once more.

The cool pine forests of western Arizona were long gone, and the Mojave Desert was even more austere than west Texas. At least west Texas had an occasional, stunted mesquite tree. The only thing the Mojave Desert had was Joshua Trees and greasewood and lots of not much of anything in between. The Joshua trees were reported to grow only in the Holy Land and the Mojave Desert—Heaven and Hell, some say. Mormon pioneers named the species for the biblical figure Joshua, because the trees' uplifted limbs reminded them of Joshua praying and pointing to the heavens. John Fremont, early explorer and promoter of the West, called it "the most repulsive tree in the vegetable kingdom." A yucca-like plant, gaunt and angular-shaped, they looked something like an old scarecrow that had been caught in a Kansas windstorm. The only greenery were long, thick, narrow, dagger like leaves with sharp, spiked tips.

Joshua Tree in bloom

Greasewoods are thorny, many branched shrubs that grow three to seven feet tall. The bark is white or dull gray in color. Leaves are fleshy, almost succulent, and grow from one to two inches long. Some animals forage on the plants, and they provide shelter to small animals.

The hot, July sun blazed down through a cloudless sky, sending even the hardy desert animals into the welcome shade of the sparse greasewood. I didn't know all of this as I drove through the Mojave Desert the first time—all I knew was that the sun was blindingly bright and hot, and I wished I had air conditioning.

Since it was July 4th the base was essentially closed, but Johnny Armstrong, whom I had met and worked with at General Dynamics on the B-58, had arranged for a place for me to sleep in the BOQ that first night. I had, at last, arrived at the one job in the entire Air Force that I really wanted.

As an aeronautical engineer and an airplane enthusiast, I knew quite a bit about Edwards AFB, and had learned even more in anticipation of my assignment there.

In 1882, the Santa Fe Railroad ran a line westward out of Barstow toward Mojave, California. A water stop was located, at the edge of an immense dry lake roughly 20 miles southeast of Mojave. The water stop was known simply as "Rod" because the lakebed was then called Rodriguez Dry Lake. By the early years of the new century, "Rodriguez" had been anglicized and shortened to "Rogers." First formed in the Pleistocene Epoch, and featuring an extremely flat, smooth and concrete-like surface, Rogers Dry Lake is a playa, or pluvial lake that spreads out over 44 square miles, making it the largest such geological formation in the world. Its parched clay and silt surface undergoes a timeless cycle of renewal each year, as water from winter rains is swept back and forth by desert winds, smoothing it out to an almost glass-like flatness.

In 1910, Clifford and Effie Corum settled at the edge of this lakebed. In addition to raising alfalfa and turkeys, the enterprising Corums located other homesteaders in the area for a fee of $1.00 an acre and, as those settlers moved in, the Corum brothers got contracts for drilling their water wells and clearing their land. They also opened a general store and post office. Their request to have the post office stop named "Corum," after themselves, was disallowed because there was already a "Coram," California. So they simply reversed the spelling of their name and came up with "Muroc."

Before World War II, the Army Air Corps conducted its flight tests at Wright-Patterson AFB in Ohio, near Dayton, not coincidentally the home of the Wright Brothers. After the war, the Army Air Corps decided to move its main flight test operations to the Mojave Desert in order to benefit from the better flying weather. Whereas Ohio has lots of rain and snow in the winter, the high desert just north of Los Angeles, California, has about 360 flying days a year. Also, because of the remote location and its over 360,000 acres, secret tests of new military aircraft and weapons systems could be done better at Muroc Field. And, let's face it, when some of these new airplanes fell out of the sky there was less chance they would hit anyone.

A small cadre of test pilots, test engineers, and support people moved form Dayton, Ohio, to Muroc, and set up the new flight test center.

Here was where the first American jet plane, the P-59 made its first flights in October 1942. So secret that when trucked to Muroc Army

Air Field the Army Air Corps mounted a fake wooden propeller on its nose. There was a story that some Army Air Corps pilots training in California reported seeing this strange plane flying with no propeller. The XP-59 pilots then hatched a plan to discredit those

XP-59 with fake wooden propeller

rumors. One of them donned a guerrilla suit and stuck a big cigar in his mouth. Then when pilots reported seeing a guerrilla flying an airplane with no propeller and smoking a cigar, no one believed them, and our new jet plane's secret was safe.

Here in October 1947 Chuck Yeager became the first man to "break the sound barrier" in the orange Bell X-1. This was one of the few test flight in which no one really knew what would happen. Mach 1.0, the speed of sound (Mach number represents the ratio of the airplane's speed to the speed of sound, 763 miles per hour at sea level), truly represented a predicted "barrier" at that time. All of the empirical aerodynamic equations that accurately predicted airplane performance predicted that the aerodynamic drag would become infinite at Mach 1, the speed of sound. A few World War II planes that had ventured too close to Mach 1 had gone out of control and disintegrated. But a .50 caliber bullet was known to travel faster than the speed of sound, so Bell Aerospace designed a little airplane shaped like a .50 caliber bullet, and stuck a rocket engine in the tail. Because of fuel limitation, the tiny plane was carried aloft under a Boeing B-29 Super Fortress and dropped. The rocket engine was ignited, and the plane accelerated to see what would happen. After the fuel was burned, the plane would glide, dead stick, back to the dry lakebed at Muroc. The Bell test pilot wanted too much money for this dangerous first flight, so Chuck Yeager agreed to do the tests for his meager Air Corps salary, and flew "Glamorous Glennis," named for his wife. When the people on the ground heard the first sonic boom, they thought he had exploded, but it was just the sound of Chuck Yeager flying into history.

Boeing B-29 Super Fortress with Bell X-1

Bell X-1 in flight

Many other exciting planes flew at Muroc, including enhanced version of the X-1: the X-1A, X-2, X-3, and so on, rapidly expanding the flight speeds to Mach 2.4 in the Bell X-2. The Douglas X-3 was designed to fly at Mach 3, but was so seriously underpowered that it could barely attain Mach 1 in level flight.

Bell X-1E

Bell X-2

Douglas X-3

But arguably the most notable airplanes to fly at Muroc in its early days were the Northrop XB-35 and XB-49 flying wings.

Jack Northrop's vision was to remove all parts of the airplane that contributed to drag and build an airplane that consisted only of the parts that contributed to lift and payload. His first attempt was the XB-35, which was only a wing, with a cockpit in the center leading edge, and four pusher propellers on the wing trailing edge. This plane was slow, and the Army Air Corps wanted jets, so Jack built another, larger wing, and put eight jet engines on the trailing edge of the wing to become the XB-49.

Northrop XB-35 Flying Wing

The XB-49 was far ahead of its time in that such a configuration is dynamically unstable about its pitch axis, even though a reflexed wing trailing edge will generate pitching moments that do not change with angle of attack. You can see this reflexed trailing edge at the wing tip in the photo of the XB-35.

Northrop XB-49 Flying Wing

The XB-49 pitched up during a stall test and crashed, killing its Air Corps flight test crew including its pilot Captain Glenn Edwards. Muroc Army Air Field was then renamed Edwards Air Force Base, also in honor of the fledgling U.S Air Force's new autonomy as a full service organization. Basic research and program management activities were still conducted at Wright Field and Patterson Air Force Base, but the vast majority of Air Force flight testing was transferred to Edwards.

Northrop's XB-45 lost out to General Dynamics' awesome 10-engine B-36 *Peace Keeper*.

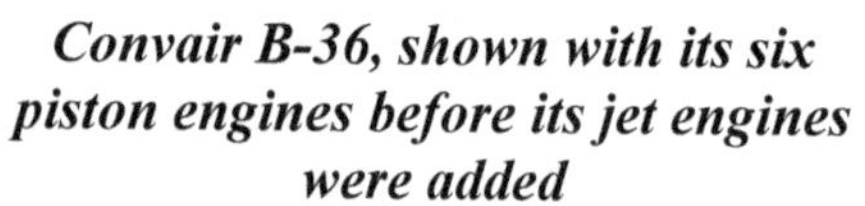

Convair B-36, shown with its six piston engines before its jet engines were added

Convair B-36 with two jet engines under each wing tip

It was not until 40 years later that small, sophisticated computers could solve the instability problem and bring Jack Northrop's foresight to fruition in the incredible B-2 *Spirit* bomber.

Northrop B-2 **Spirit** ***Bomber***

The Navy did its flight testing at Patuxent River Naval Air Station, Maryland, on the Chesapeake Bay, and there were numerous other military test bases for missile, engine, nuclear, and electronics testing, but Edwards, as Thomas Wolfe wrote in his book *The Right Stuff*, was the pinnacle of the pyramid, the ultimate test base, and Chuck Yeager was sitting right on top at the peak of the pyramid! And here was where the B-58 would be flight tested, whose first flight I had witnessed the previous November when I worked at General Dynamics. I could actually fly on this incredible airplane!

With all of these exciting airplane tests going on I was delighted to become a part of it. For an aeronautical engineer interested in flight test, Edwards was the "pot of gold at the end of the rainbow," the ultimate dream job, a flight test engineer's most ardent wish come true. Our performance, and stability and control flight tests would prove whether the airplane contractors had met their contract performance guarantees, and would get paid millions of dollars. Our tests would provide the detailed data for the pilots' flight manuals so they could plan and fly their missions with confidence and safety. Our operational suitability flight tests would demonstrate how the airplanes could fly those missions most effectively. My excitement was not the kind that pounds your heart, catches your breath, and makes you jump up and down on the bed; but the kind that wakes you up early every morning with a big, silly grin on your face and rushes you to work. The kind that brings the days end much too soon. The kind that makes you reluctant to take vacations because you don't want to miss one single, exciting day. The kind of excitement that occupies your every waking moment and fills every conversation with the day's exciting events and your plans for the morrow, and with the knowledge that what you are doing is important to prove to the Air Force that the planes can fly their assigned Air Force missions successfully to defend our country, and can bring their crews home safely to their families.

And, most of all, it was all really neat stuff! How many people can say that their job is not only so important, but so much fun?

So, eagerly, I processed into the Air Force the next day, and reported for duty in Flight Test Engineering. But first I had to report in to the Edwards Base Commander. I presented myself to Colonel Meyers' secretary, told her that I was reporting for active duty, and she announced me. Upon her invitation to enter his office, I strode up to his desk, came to straight attention, clicked my heels smartly, and announced: "Second Lieutenant Robin K. Ransone reporting for duty SIR!"

2nd Lieutenant Robin K. (Rob) Ransone

Col. Meyers returned my salute, glared at me, and asked if I had a light for his cigarette. I said "Yes, Sir," and offered him my lighter—almost everyone smoked in those days.

He remarked "The only thing that Second Lieutenants are good for is lighting senior officers' cigarettes!"

"Yes, sir." I had learned at A&M not to take such hazing personally. There are times when military duty is like playing a game: you play the game, and all goes smoothly. Don't make more trouble for yourself when you meet someone like this by arguing, trying to stand up for yourself, or making smart retorts. Fortunately, there are very few such officers. The rest of the officers at Edwards, including all three commanding generals that I served under, were very personable and civil, and respected what we junior officers contributed.

I finished this interview, completed my physical examination for entry into active duty, and reported to my work assignment in Building 1400, Flight Test Engineering Directorate. I was assigned to the Fighter Division.

When I first arrived at Edwards it was a place where flight test success or failure depended more upon the quality, capability, and enthusiasm of the individual—yes, and luck—rather than upon leveling regulations and rules that claimed to legislate order, but really usurped initiative and innovation. "Ad Inexplorata," the center's motto: "Towards The Unknown" included searching out new test approaches and data analysis innovations, as well as demonstrating new aircraft concepts.

Indeed, we flight test engineers did more than collect and analyze data and write reports. We were multi-dimensional—knowing why the tests were needed so their results could be made most useful; how to make the tests safer; and how to apply the lessons learned, not only for the next test, but also to the betterment of the next generation of technology.

Later, when we became more organized and disciplined, there were probably more hours of boredom and fewer moments of stark terror. In an over-structured environment where forgiveness was easier to get than permission, there was less tragedy, but there was also less fun. And in those days when the hottest Air Force airplanes were the

F-86, F-100 and F102, and the fastest could go only to Mach 1.2—only slightly faster than the speed of sound—flight testing was really fun! We manned our flight stations, our photo panels, our oscillographs, our slide rules, and our desks, and worked wonders one day at a time. We planned when we could, but sometimes we just had to make it up as we went along.

The following stories relate some of my more memorable moments from the eleven wonderful, exciting, and rewarding years that I spent at Edwards, both as an Air Force officer and as a civilian flight test project engineer.

In much later years I especially appreciated a line from the Hollywood movie *Raiders Of The Lost Ark*, from Harrison Ford to his girl friend, Karen Allen. They had just escaped from the Nazi commander's tent, Ford had set fire to most of the Nazi encampment, the Nazis were looking all over for them and shooting at everything that moved. Allen turned to Ford and demanded: "What's your plan?"

Ford looks at her in amazement; "Plan? *Plan?*" He countered. "I *have no plan* – I'm making it up as I go along!"

And, many times, so did we.

Life As A Young, Broke, Second Lieutenant

Everything was a lot cheaper in 1957 than it is now. Our 2nd Lieutenant base pay was $227.00 a month, plus $45.00 a month "subsistence," to be used for food. Flight pay for rated 2nd Lieutenant pilots was an additional $150.00 a month, but since we engineers were not rated, we got only $110.00 a month "hazardous duty" pay. For this, we had to fly at least four hours a month, which could be attained if we flew 12 hours every three months. When we were actually flying test missions, and there was a seat in the aircraft for us, there was no problem getting our flight time, but after the flying was done, and the real work of data analysis and report writing began, sometimes it was difficult to get our time. Edwards was probably the only place in the Air Force where a C-47 transport would fly locally with six flight test engineers, four navigators, and three pilots, all earning their fight pay

To put our pay—puny though it was—into perspective: a brand new Buick sedan cost only about $ 3,500.00, and a new Rolls Royce cost only about $ 16,000.00. So we could actually "survive" on our meager pay.

But it was still a very tiny income compared to my previous salary at General Dynamics. We had been working 60-hour weeks, with time and a half for over time and sometimes double time! In fact, the *income tax* that I paid on my salary with overtime when I worked at General Dynamics between January through June was *more* than my Air Force salary for July through December of that year!

I had paid my own salary! What a *bargain* the Air Force got that year!

So my total pay, as a 2nd Lieutenant, was $382.00 a month, of which only the $227.00 was taxable. We got free room in the Bachelor Officers' Quarters (BOQ), which included heating, air conditioning, and maid service, and there were washing machines and clothes dryers where we did our khaki uniforms. We all had irons and ironing boards, and got pretty good at ironing in military creases. We had to pay for our meals, of course. The good news was that the Officers' Club was right next to the BOQ—the bad news was that we couldn't afford to eat dinner there every night. Normally, we just had lunch at the O Club. In addition, I was paying for two new Buicks—my 1956 Century and Mother's 1955 Super—for a total monthly car payment of about $150.00.

Each BOQ room had a refrigerator, so we had cereal for breakfast. We discovered TV dinners at that time. They were not nearly as good as they are now, nor was there as much variety, but we made do. We each had a private room, but shared a bath room with one other junior officer.

Fortunately the water in the lavatory was very hot, so we would crimp the aluminum foil down securely around the edges of the TV dinner and set it in the sink, running hot water over it for about a half hour. Usually this worked fine, and we could enjoy a lukewarm dinner. But sometimes there was a small pinhole in the aluminum foil, and we ended up with meatloaf soup, fried chicken soup, or Salisbury steak soup.

Our big break came when we discovered little, electric toaster ovens. These had a heating element in the top, and were just big enough to hold a TV dinner tray. We attached an aluminum sheet on the front to contain any splattering, and could now really heat our meals. We could even buy a piece of round steak, cut it into four pieces, and grill it in these little ovens.

Man! We were really living!

To give you an idea of the weird things had kept us young, arrogant, aeronautical engineers entertained, consider our reaction to watching the rocket engine tests across the air base. The BOQ was located on top of a low hill, and looked across most of Edwards AFB, the runways, and about fifteen miles of Rogers Dry Lake, to Leuhman Ridge. The Rocket Test Site was built along the top of Leuhman Ridge, on the far east side of Rogers Dry Lake. The rocket engine test stands hung over the edge of the ridge so they could fire straight down, as they would in an actual rocket. The sides of the test area were heavily concreted and there were huge water washes that cooled the concrete when the engines fired. The engines were monitored and controlled from solid concrete blockhouses that protected the test engineers in case of an explosion. In 1957, during the early days of rocket engine development, explosions were not uncommon, but only occasionally did they destroy a test stand. The rocket propellant was usually unsymmetrical dimethylhydrazine, and the oxidizer was red fuming nitric acid or hydrogen peroxide—not the puny hydrogen peroxide you buy at the drug store, but really potent stuff that you didn't want to get on you! The ideal rocket fuel and oxidizer was known to be liquid hydrogen and liquid oxygen, but no one then knew how to build rocket engine nozzles that could stand the extreme temperatures of such a combination. Now rocket scientists know how to do that. Neil Armstrong could not have flown to the moon on hydrogen peroxide and unsymmetrical dimethylhydrazine. His Rocketdyne F-1 engines burned liquid oxygen and kerosene.

Rocket Test Laboratory, Leuhman Ridge, with Edwards AFB in the background

At that time the Rocket Test Site was testing the rocket engines for the Atlas and Titan Intercontinental Ballistic Missiles (ICBMs). From the BOQ we could see the fire and hear the engine's roar clear across the desert. Since sound didn't travel as fast as light, we saw the fire long before we heard the sound, and the sound persisted long after the fire went out.

We would time how long we heard this sound after the fire went out, and do our engineering calculations. Since we knew how far away the rocket test stands were from the BOQ, and how the speed of sound varied with air temperature, we would calculate the air temperature from these times.

You have be an engineer to really appreciate this bid of foolishness!

The Air Force decided that its officers didn't get enough exercise and were getting out of shape, so they decreed that each year we must pass a rigorous physical examination that included a number of pushups, chinning, "jumping jacks," and running. For running, the instructor set up two pylons in the parking lot, and we had to run, in pairs, around these two pylons a certain number of laps within a prescribed time.

Running full tilt around these pylons on a flat, asphalt parking lot surface, was a bit hard on the ankles, so the second year I tried something different. Being an engineer, I realized that regardless of how I did it, I had to stop my northern speed and accelerate back to the south after passing the pylon, and then reverse the process at the southern pylon. I calculated that it took the same amount of energy to reverse my direction, but that it would be easier on my ankles and I would actually have to run a shorter distance if I just ran up to the pylon, stopped, stepped around the pylon, and then sprinted back in the other direction.

So that is what I did.

I would run right up the pylon, stop, step around it, and run back to the other pylon. Run, stop, step. Run, stop, step. Back and forth.

This drove my partner nuts!

"What the *hell* are you *doing*??" He wanted to know.

It didn't make him feel any better when I finished my run about three laps ahead of him!

Another interesting night sight from the BOQ was an occasional bright glow to the northeast from the distant Nevada nuclear test site when a thermonuclear bomb was detonated. This was at the height of the "Cold War" with the Communist countries, and things sometimes got very tense.

A not so interesting sight occurred about once a year. We would notice a bright orange glow to the south, towards Los Angeles. When we saw this glow in the night sky we knew that another forest fire was ravaging the Angeles National Forest, and the Angeles Highway would be closed through the mountains until the Smoke Jumpers had extinguished the blaze.

Many of us built control-line, gas engine-powered, model airplanes and flew them off the hard, smooth surface of Rosamond Dry Lake. Johnny and I had great fun with these. Johnny had gotten me interested in control line models when we were both working at General Dynamics before coming to Edwards. Radio-controlled model airplanes were not as popular as control-line because of the (then) bulky and expensive radio equipment. In control-line model planes, there was a small bell crank in the center of the model with two thin steel control lines connected to a handle that the "pilot" held in the center of a circle. When the pilot pulled each line by moving the handle, it moved the bell crank in the plane and moved the elevator up or down. Centrifugal force held the plane in the circle, and the pilot turned around and around as the plane circled him. Normally, the circle radius (and control line lengths) were on the order of 50 feet or so, but with the plane flying at 50 or 60 miles per hour, the pilot rotated in the center of the circle at a fairly high rate —it took a few flights to get over being dizzy. With only up and down motion available, no rudder or aileron controls were needed.

Since I loved the B-58, and planned to be a flight test engineer on the stability and control flight tests at Edwards, I built a scale model of the airplane from plans that I got from General Dynamics. I also designed and built a retractable landing gear. It was ingenious, with full retracting gear and wheel well fairings, all powered by a small electric motor. With the motor and batteries in the model, it was pretty heavy for a control line model airplane. I had everything I needed except for a small shaft and crank to convert the rotary electric motor output to the up and down motion needed to retract and then extend the landing gear. Roy Pieberman ("The Pebe") came to my rescue. He was assigned to the Power Plant Branch, and had his shop machine

Johnny Armstrong and me (on right) with delta wing control line model airplanes

the small aluminum part from my drawing. When it was done, he telephoned my office with the curt and strange message: "Six six zero is no longer AOCP!" [Translation: "660" was my model airplane's tail number, same as the number one B-58, and "AOCP" was the Air Force term for "out of commission for parts.")

With this last part installed, and using the two steel control lines to connect to a switch on my control line handle, the landing gear cycled perfectly!

The big day: First test flight!

Trouble. For some reason the landing gear jammed and I could not retract it. Oh well, not to worry—we'll just leave the gear down for the first test flight.

Engine start, Johnny as crew chief, holding the plane until I was ready. I gave the signal, and he released the plane and it roared across the dry lake bed and climbed steeply. It had enough power, even at its heavy weight. I made the usual flight maneuvers: steep climbs, dives, wingovers (a maneuver where the plane climbs sharply up from a low altitude, flies directly over the top of the circle, dives down the other side, and pulls out sharply just above the ground. It was a great flight—until the plane ran out of gas and the engine quit.

I was never very good at "dead-stick" landings.

Since the B-58 had a relatively small, triangular-shaped "delta" wing, and my model was pretty heavy, it came down hard. A very hard landing, in fact.

The point of the matter is—the airplane hit so hard it broke in half!

But Johnny and I were not injured (ha ha).

I removed the gas engine and electric motor, poured gasoline on it, set fire to it, and took pictures. Flight testing is a dangerous business, but "somebody's got to do it!"

The dry lakes at Edwards were one reason that the Air Force Flight Test Center was established there in the late 1940s. With over fifteen miles of hard surface, on which planes could land in any direction, it was also a popular alternate for safely landing commercial airliners or military airplanes with landing gear or brake problems.

If the winter and early spring rains were sufficient, two amazing transformations took place. Tiny, prehistoric shrimp would hatch out of the hard, dry lakebed clay, wiggle around, reproduce, and die. And the entire desert would blossom—California poppies, and wild flowers of every kind and color, including the rare Desert Candles. The desert was covered in a kaleidoscope of color as far as the eye could see. It was beautiful.

California Poppies

Desert Candles

It also proved how resourceful nature can be, because neither the shrimp would hatch nor the flowers would germinate unless there was enough water for them to compete their full reproduction cycle, ready to reappear with the next big rain, which might be several years away.

A popular pastime in the BOQ was building hi-fi kits. We couldn't afford ready-made hi-fidelity amplifiers and fancy speakers on our meager junior officers' pay, so we bought our components in kit form. The most popular was from Heathkits. The price was less than half the price of ready-made components, and the kits came with excellent and easy to follow instructions. There were pictures of the chassis, controls, sockets, and tubes—this was before everything was transistorized—and the tubes were usually from England. Each connection was numbered, and the instructions would direct us to "[ ] Connect a 20 K-Ohm resister (red, yellow, red) between S2 and C4 (NS)" The "S2 and C4" indicated the connection points on the sockets and fittings, and the “NS” meant “don’t solder yet.” Once we had completed that step, we checked the brackets at the front of the instruction.

We normally bought separate preamplifiers from the amplifiers because the performance was better on the vacuum tube-type systems. The alternating current used to heat the vacuum tube filaments created a 60-cycle hum in the playback. Later, with the use of transistors and integrated circuits, separating the pre-amp from the amp was not necessary. A big improvement was when Dyna-Kit introduced a pre-amp that had direct current on the filaments so there was no hum at all.

Whenever one of us got a new component—pre-amp, amp, turntable, speaker—we would all pool the best of our equipment, connect the new component, and try it out. We always used my Antal Duratti and the Minneapolis Symphony Orchestra recording of Tchaikovsky's wonderful *1812 Overture*, with the cannons and the bells at the end. This really gave the new system a workout! We would stand in front of the speakers and direct the orchestra with a stick, bringing in the various instruments, cannons, and bells.

Although there were many popular recordings at the time, we concentrated almost exclusively on classical music, probably because a full orchestra’s performance really showed our hi-fis to their best advantages. In retrospect, I think it also indicated something of the caliber of the AFFTC junior officers—we were not only engineers, but were perhaps “both-brained” as well, exercising both our creative and logical perspectives.

One fellow flight test engineer once told me that I was the only person he had ever met who walked down the hall whistling the Gregg Piano Concerto…

With our Aeronautical Engineer classifications, we in Flight Test had few actual military duties. An occasional “Report of Survey” (where we would spend about a day investigating the accidental destruction of an Airman’s issued property), helping an accountant count the money at the base Disbursing Office to ensure that all the accounting records were accurate, or standing Officer of the Day duty about twice a year.

Counting the money provided a sense of power: since the inspection was a surprise, we would just walk up to the disbursement window and announce that the office was closed for the rest of the day, and then we would go into the drawers and vault and count the money under the watchful eyes of a bona fide accountant.

Officer of the Day was also interesting. As such, we represented the Center Commander. We would report to the Base Headquarters, be issued our official Sam Brown Belt, and receive any standing orders or information on anything unusual expected for the evening. Our duty tour for weekdays was from 5:00 PM that evening until 7:00 AM the next morning. Actually there was also a Sergeant of the Guard who did most of the work, and had to stay in the office overnight, for which a cot was provided. We officers normally went back to our BOQ rooms (or base housing for the married officers) about 10:00 PM, leaving our telephone numbers with the Sergeant in case anything unusual came up. The next morning, we would brief the morning staff on anything notable, and make appropriate entries in the log.

The only really unusual thing that happened while I was at Edwards was one night, about 2:00 AM, the Sergeant of the Guard telephoned the Officer of the Day to report that a small, empty building on the old South Base area had burned. The OD asked whether there was anything left of the little building, and the sergeant answered "No, it burned completely to the ground." "Then there is really nothing for me to see," reasoned the OD, and he went back to bed.

The most embarrassing thing to happen to an OD was during Armed Forces Day, when the base was open to thousands of visitors to view the many interesting (unclassified) airplanes and watch an air show with many of them flying. The flying also included North American Aviation's remarkable Bob Hoover, flying incredible acrobatics in a North American F-86 fighter or his own beautiful, yellow P-51 Mustang. There were also other aerobatic exhibits and the grand finale was always the USAF Thunderbirds aerobatic team.

On this particular day, however, the OD was a rather inexperienced Second Lieutenant who was not very familiar with some of the plumbing in hanger areas. Dressed in his best Class A uniform and grandly decked out in his official Sam Brown Belt and official

"Officer of the Day" arm band, he marched proudly into the men's room. There, in the middle of the room, was a large, granite, bowl-shaped fixture. It was about six feet in diameter, about eight inches deep, with a metal ring around the base and a pipe sticking up in the middle.

He thought this must be the place, since it was the right height, so he strode confidently up to it and unzipped his pants.

Just then, however, to his great amazement and embarrassment, another officer walked up to the fixture and stepped on the metal ring. This caused a cascade of water to spray from the center pipe, all around the bowl, and the other officer proceeded to wash his hands and glare accusingly at the OD. The red-faced OD zipped up and made a hasty retreat!

Not all unusual events at Edwards were so funny. During those early days, during which jet engines and advanced flight control systems were not as reliable as they are now, there were, unfortunately, many accidents at the Flight Test Center. It was said that flight testing was "hours and hours of boredom and brief moments of stark terror!" Most of the time the "brief moments of stark terror" ended happily, with important lessons learned and great stories for cocktail time. Sometimes, however, they ended in tragedy.

Missing Man Formation, a final salute

It was a somber moment when a flight of planes would fly low and slow down the flight line in the "Missing Man Formation." This was usually an incomplete diamond formation, with the missing plane representing our fallen comrade. Sad as this was, we had to accept it and continue. We all knew the risks, but our training, professionalism—and most of all, our attitudes—made us believe that it probably wouldn't happen to us.

Our work was critically important. We were testing the Air Force's very latest aircraft, and our job was to find out what they could do and what they couldn't do so that the aircraft could safely fly their intended missions. The aircraft contractor would fly the first flights to prove its airworthiness, and then we would fly performance tests to gather the data necessary for the Flight Handbooks, and stability and control tests to determine the aircraft's handling and flight characteristics, and control responses. We tested against the contractual specifications and the results of our tests determined whether the airplane contractor got paid.

I remember having butterflies before some flights, but I was never scared once in the air, even during emergencies. During our occasional in-flight emergencies we were always so absorbed with solving the problem and getting the aircraft down safely, that we seldom experienced fear or thought of our own safety. We calmly analyzed the situation, assessed possible solutions based upon our knowledge of the airplane systems and performance and control capabilities, and acted accordingly. Sometimes we had several minutes to do this, but sometimes we had to do it in only seconds. Those who died usually waited too long to eject, rather than ejecting too early.

Not all of the in-flight emergencies occurred in test aircraft. For example, one day Johnny Armstrong and I were getting our four-hour flight time on an ancient World War II C-54, propeller transport. The pilots were "shooting touch and goes" (repeated landings and takeoffs for proficiency training). Johnny and I were absorbed in our chess game, but suddenly realized that we had touched but hadn't gone! We had stopped on the runway. We looked out to see the fire trucks all around the plane because we had had an engine fire right outside our window.

We were so engrossed in our chess game that we hadn't noticed the engine fire.

My First Flight Test Program

My first assignment at Edwards was Assistant Flight Test Engineer on a new version of Convair's F-102 delta-wing jet fighter. Convair had discovered, through its research, that if the leading edge of the wing was curved down a little bit near the wing tips, that it reduced the aerodynamic drag and the airplane could fly farther on a pound of fuel. Thus, the F-102 Case XX Wing program was born. We tested it at Edwards.

F-102 Case XX Wing – My first flight test program with Gordon Cooper and Bob White

The Project Pilot was Captain Bob White, who later flew the rocket-powered X-15 to 50 miles altitude (over 260,000 feet) and earned his Astronaut wings. The Project Engineer was Captain Gordon Cooper, who later became one of the seven Project Mercury astronauts.

I learned a lot from this program, mostly about how we did things at Edwards. What fun! And I actually got paid for doing this—not much, but enough.

Basically, we worked as a test team, comprising a project test pilot, a project test engineer, instrumentation support, and aircraft maintenance personnel. When a new plane was coming to Edwards for testing, the project pilot and engineer would get together and write the test plan, and it would be reviewed and approved by our respective Flight Test Operations and Flight Test engineering bosses. We would also define any special test instrumentation that we would need to record the test data.

Normally the airplane contractor conducted the first, Category I, flight tests on a new airplane, demonstrating its safety and flight worthiness, which we monitored. Sometimes our pilot would fly on the contractors' tests. Then the plane would come to Edwards, where we would conduct the Category II Performance tests (to collect the data for the Pilot's Manual), and Stability and Control tests (to assess the airplane's handling qualities). The Category II tests confirmed (or identified failures) of the contractor's contractual compliance with the development specifications.

Although very advanced digital and telemetry flight test instrumentation became available by the time I left in 1968—to the point where the data were already being compiled even as the test airplane landed—in 1957 we utilized photo panels and oscillographs.

The photo panels consisted of conventional, round, instruments normally seen on cockpit instrument panels, with special lighting and a 35mm film strip camera to record the airplanes performance. These instruments were especially calibrated in the Instrumentation Laboratory so that we could make the necessary instrument corrections to ensure accuracy. Normally these instruments were for airspeed, altitude, normal acceleration, and fuel quantity or fuel used.

We read these instruments from the 35mm film strips, displayed on a big viewer, and made the corrections to ridiculous detail! For example, we would note an altimeter reading of 25,215 feet, which could only be read to about 50 feet or so, apply a corrections of 2 ½ feet for the calibrated instrument errors, and come up with a corrected altitude to put into our data analysis computations of 25,217.5 feet.

But we engineers refused to give up *any* accuracy, regardless of how vague it really was.

For stability and control tests, in addition to the photo panel, we also had an oscillograph to record the aircraft motions; roll, pitch, and yaw angles, and rates of attack, pitch, yaw, and sideslip. The data were recorded as wiggly lines on an 11-inch wide roll of photographic paper that had to be developed in the photo laboratory before we could use it. Imagine a roll of paper, fifty feet long, with a dozen or so wiggly lines on it, and a time scale that was presented in binary code. This code was little "blips" along the bottom of the paper, and the various lines of data were identified by the location of a small break in the line, referenced to the timing line. We used a ruler to measure the lines' variation relative to a datum line on the paper, and calculated the time from the binary code.

We were issued 20-inch slide rules that we used for our calculations, and we had pre-printed forms onto which we entered the raw data and then made our calculations.

Very laborious.

Everything had to be calibrated, however, which was very interesting.

For example, to calibrate the fuel tanks, we used the remarkable scales in the Weight and Balance Hanger. This was a knife-edge, mechanical balance system, which weighed each wheel individually. The airplane was drained of all fuel, the tanks purged, and the plane brought into the hanger. The hanger doors were closed so that no breeze would affect the scales' reading. After everything had settled down, the operator set the weight arm, in a chamber below the hanger floor, in order to balance the arms, and squeezed a lever that imprinted the weight on a slip of paper.

Measured increments of fuel were added to the airplane, and the weights were recorded following the same laborious procedure. Since each wheel was weighed separately, we not only got the accurate weights to calibrate the airplane's fuel gauges, but we also could determine the airplane's center of gravity.

These scales were incredibly accurate! I have seen them weighing a 160,000 pound B-58, and when the cockpit canopy was lowered, the nose wheel reaction increased by 5 pounds, and the two main landing gear reactions decreased by 2 ½ pound each!

This accuracy didn't come easily. Each year a representative from the National Bureau of Standards came to Edwards to calibrate the scales. Stored under wraps in the hanger were a number of 10,000 pound weights with small compartments in the top for BBs—the tiny lead shot used in shotguns and BB guns! The Bureau representative brought a calibrated 10,000 pound weight that he used to recalibrate each of the local weights before using them to calibrate the scales.

One of the most exciting calibrations were for the airspeed and altitude instruments. On test aircraft you may have noticed a long boom sticking out of the nose of the aircraft. This is for the airspeed and altitude instrument pickups. By placing these pickups on the end of a long boom on the airplane's nose, their readings are very accurate because the airflow they measure is not influenced by the shape of the airplane. We had two airspeed calibration aircraft at Edwards: an F-100 for high speed chase (later an F-104), and a T-37 for slow speed chase.

T-37 low speed airspeed calibration and chase aircraft

These chase aircraft had to be calibrated, of course, and we did this using a technique called a "tower flyby." For a tower flyby, the aircraft flies about 50 feet above the ground past a special tower. In the tower there was an eyepiece that swung down from the ceiling on a long arm. On the window were scribed horizontal lines. By looking through the eyepiece and noting the line on the window along which the airplane appeared to be flying, the precise height above the runway could be determined. By comparing this height with the aircraft's altimeter reading, the airplane's precise airspeed indicator and its altimeter could be calibrated.

How can you correct its airspeed from its altitude? The simple answer is, "Don't ask! Just trust me—it works!"

Another one of the first tests in which I participated at Edwards was one of these "tower flybys" to calibrate the airspeed and altitude instruments on a North American F-100 fighter from the Navy's China Lake flight test facility, about 75 miles north of Edward.

F-100* Super Sabres *in Flight

These tower flybys were normally done first thing in the morning when there was no wind to corrupt the data. On this particular morning, Gordon Cooper and I were in the tower. This tower was located at the old, otherwise deserted part of Edwards known as South Base, which had been the original facility before the base was expanded and new runways and hangers constructed.

It so happened that this morning workmen were on the runway installing special equipment for forthcoming aircraft arresting barrier tests—equipment that someday could safely stop a landing aircraft whose brakes had failed.

It was just after daybreak. They had their backs to the bright, rising sun, of course. The pilot wanted the sun at his back, too, because he was flying at very high speed only 50 feet above a concrete runway.

His first pass was at Mach .95, 705 mph, just under the speed of sound, full afterburner! At that speed we could not hear him coming.

Neither could the workmen.

When he was directly in front of us, complete silence—then the sound hit us!

The first indication they had of any airplane around was when the plane flashed above them at only 50 feet, and the thunderous roar of the afterburner hit them!

I never saw anyone run for cover so fast in my life! By then, of course, it was too late. The plane was already gone, and climbing and banking around for his next pass.

When they finally had the nerve to get back on the runway, they faced the sun, so they could see the plane coming.

What fun!

My First Promotion

I had been on active duty at Edwards for only six months when I became eligible for promotion to 1st Lieutenant. Johnny Armstrong also became eligible the same month—February, 1958. It was then that a strange set of circumstances developed regarding our promotions.

At that time, 2nd Lieutenants became eligible for promotion if it was two years from date of commissioning (i.e., graduation with an ROTC commission) with at least six months of active duty. We were also eligible if we had eighteen months active duty. Both Johnny and I had two years from date of commissioning. Johnny had eighteen months active duty, but I had only six months active duty. He was on active duty a year longer than I.

But I had gotten commissioning moved up a week at A&M so that we could have a graduation exercise. This made my date of commissioning a week before Johnny's.

So, even though Johnny had been on active duty for a year and a half, and I had been on active duty only six months, my promotion came a week before his.

I proudly bought my silver bars, and asked Johnny to pin them on for me.

Grudgingly, *very* grudgingly, he agreed. He pinned them on, and then, grimacing, made some snide remark.

I looked him straight in the eye, and with as serious a tone as I could muster, exclaimed: "If there's *anything I can't stand*, it's a *smart-ass Second Lieutenant*!"

Fortunately for me, Johnny has a great sense of humor, and we were both laughing so hard that he didn't hit me!

The next week, I pinned on his silver bars, with just as much pride as when I received mine from him.

Preparing for High Altitude, Supersonic Flight Testing

Having worked on the B-58 performance predictions and witnessed its first flights at General Dynamics Corporation in Fort Worth, I was excited to be assigned to the B-58 flight test program when it came to Edwards. But before I could fly on this magnificent airplane I had to take special training for high altitude flight. This included a special flight physical examination and certification by the Flight Surgeon that I was physically fit for high altitude flight.

My flight physical was very thorough, and included fingerprinting and recording dental records for identifying me should I be killed in an accident. My foot prints were also taken. Surrounded by 95,000 pounds of jet fuel in the B-58, the protection of my heavy flight boots might make my footprints the only means of identifying what might be left of me.

Now, that's a sobering thought!

As a human flies higher and higher, the air becomes less dense, and breathing becomes more difficult. There is the same *percentage* of oxygen in the air, but since the *air pressure* decreases with increasing altitude the *partial pressure* of the oxygen in the air becomes too low to force it through the membranes in the lungs, and the person suffers "hypoxia," or lack of oxygen. At a certain altitude, the person will "black out," lose consciousness, and, if not given oxygen immediately, will die. There is no problem up to an altitude of about 10,000 to 12,000 feet, but above that altitude, people need either an oxygen mask or a pressurized cabin.

The B-58 flew routinely above 50,000 feet, with a pressurized cabin, so normally there was no problem. But we had to wear oxygen masks in the event of cabin pressurization failure or if we had to bail out. Cabin pressurization could be lost by a failure in the pressurization system or by a failure of a window so I had to take the altitude indoctrination training.

It was very interesting.

First, we were taught what happens to a person when he or she is deprived of oxygen, and how to recognize the symptoms of hypoxia: finger nails start turning blue, speech becomes slurred, and eye sight becomes blurry. We learned that below about 42,000 feet, we could suck in oxygen through our oxygen masks just by breathing normally—so-called “breather-demand.” Above 42,000 feet, however, the oxygen systems changed to a “pressure demand” system. This meant that the oxygen was being feed into our masks at a higher pressure. This was needed in order to force the oxygen through the membranes in our lungs and into our blood stream. If we relaxed, the mask would force oxygen into our lungs, so we had to consciously force ourselves to exhale against that pressure. Having to talk while exhaling against this pressure made it very difficult to talk. It is not something that you want to do very long! We were also taught how to equalize the air pressure in our inner ears if we had ear pain. There was no problem in going up—the air pressure equalized on both sides of our ear drums automatically. The problem sometimes occurred when coming down, and the air outside our ears was increasing in pressure. You experience this when descending in a commercial airliner. The stewardess may give you chewing gum, and babies cry to relieve the pressure. We were taught to press out fingers against our nose, close our mouths, and exhale against our blocked nose and mouth. This would “pop” our eardrums and equalize the pressure, alleviating the discomfort. This procedure is called “valsalva.”

After this indoctrination, we were issued oxygen masks, formed into teams of two, and entered the altitude chamber. The altitude chamber was a steel cylinder about ten feet in diameter and maybe 15 feet long. The outside was painted a dark blue, of course, and the inside was painted white. We donned our masks, the heavy, airtight door was closed, and a pump began to pump the air out of the chamber. There was an altimeter in the chamber, so we could see our pressure “altitude” as we “ascended.”

The chamber “leveled off” at about 25,000 feet, and the instructors asked how everyone was doing. A “thumbs up” meant “OK,” and one or more fingers held up signified that someone was uncomfortable—the more fingers, the more discomfort.

The instructor had one of us take off his mask, and start talking. After about 30 seconds, the subject’s voice became slurred, and the instructor told him to put his mask back on. The subject just kept on talking, and was not capable of helping himself. Whereupon the instructor immediately put his mask on for him, and the subject recovered immediately.

The lesson of this was that one must be able to recognize his own symptoms of hypoxia in time to take preventative measures or he will not be able to save himself, and would probably die.

With this demonstration, half of us removed our masks while our partners watched us. When we recognized the onset of hypoxia, we put our masks back on. One or two had to have their partners replace their masks.

After this demonstration, we were brought back down to "sea level," and left the chamber, qualified for high altitude flight.

Except…since the B-58 was capable of flights above 60,000 feet, I had to have further training—partial pressure suit training. Above 60,000 feet, the air pressure is so low that my blood would "boil" and I would die if the cabin pressure failed—an instantaneous case of the "bends" suffered by deep ocean divers.

My partial pressure suit was a strange looking uniform, not at all like the full pressure suits worn by the astronauts. It was a custom made, skin tight, nylon suit with one-inch tubes running down my shoulders to my wrists, and down my sides from my arm pits to my ankles. There were thin straps sewn to the suit an inch or so back from these tubes and wrapped around the tubes. Normally, there was no air pressure in these tubes, and the suit was actually very comfortable. In the event of a cabin decompression, however, the aircraft pressurization system would automatically

Me, on the right, in my partial pressure flight suit

inflate the tubes, tightening the straps, and pulling the suit tightly around me. I had a special face plate with an opening through which I could drink water through a straw, my microphone and headset were in my helmet, and there was a thin wire heating element in the faceplate to keep it from fogging up when I breathed. Our flight boots had no pressurization, but the boots were tight enough that pressurization was not needed. We had gloves with pressure tubes along the backs of the fingers, but they were unwieldy for taking notes and operating the aircraft and test equipment, so we didn't bother with the gloves.

During normal flight at high altitude, the suit was not pressurized, but in the event of a bailout, I would have to rely on the rocket-powered ejection seat to get me out of the airplane. Upon ejection, my suit would be pressurized automatically, and I had a small, high pressure oxygen bottle in the seat pack to give me air while I descended. It would be good for about 10 minutes. But since it might take me almost an hour to descend the 8 ½ miles down to 15,000 if my parachute opened at 60,000 feet, I would have to free-fall for several minutes until I got down to where I could breath without the oxygen mask. There was an altimeter in the chute pack that would automatically inflate the chute at 15,000 feet, and I was cautioned not to open the chute manually, unless I was below 15,000 feet and the chute had not opened automatically. Since it might be difficult to judge my height while free falling from 60,000 to 15,000 feet, during actual flights I always noted the altitude of the cloud cover, and therefore knew that if I entered the clouds and my chute hadn't opened yet, then I should pull the rip cord myself.

I think that would be a pretty exciting but lonely descent! Fortunately I never had to do it.

There was another problem with ejecting from the B-58—it not only flew very high, it also flew very fast! If I ejected at supersonic speeds, I would be ejected into a 1,400 mph air stream with a wind blast air pressure of 1,200 pounds per square inch! In order to keep me from flailing about, the ejection system had some restraints built in.

There were ankle straps that I attached before we took off, that were attached to the floor of my crew station. These would pull my legs together and lock them together as the seat left the airplane and the attachment to the floor would break away. There were shoulder straps that were attached to the buckle of my seat belt, and there were two loops of heavy nylon webbing attached to the buckle. If I had to eject, I would have to put my arms through those loops, then pull them tightly as I gripped the handles on each side of the seat, which I would pull to initiate ejection. These restraints, and my grip on the seat handles, secure my arms when I hit the wind stream.

This was a very complicated system, so fortunately I never had to use it. In fact, I don't know whether anyone ever used it successfully, supersonically, during the many crashes of the B-58 flight test and operations program. I'll show you pictures later in my story about *There's a brown bear in the back seat!*

But back to my altitude training.

I donned my partial pressure suit, with the help of the altitude training staff, and entered the altitude chamber with a trainer, who had only an oxygen mask. We went up to 25,000 feet and checked to see that everything was working properly. Then I went into a smaller chamber, just large enough for me and a seat, was hooked up to the pressurization system, and the trainer went back into the larger chamber, closed the heavy airtight door, descended to sea level, and exited the chamber.

Now there was only me in the small chamber, and no one in the larger chamber. I was in constant microphone contact with the trainer at all times.

The larger chamber was raised to 85,000 feet, and when I replied that I was ready, a large valve was snapped open, and I experienced an instantaneous, explosive decompression to 65,000 feet!

WOW!

Actually, I felt no change in pressure because my partial pressure suit inflated instantaneously when the chamber pressure dropped. The only sensations were the loud explosive bang as the valve opened, and the moisture in my chamber condensed and fogged the chamber.

I was brought back down to sea level, and declared ready for B-58 flights!

What fun!

B-58 Flight Thrills At 1400 mph

The B-58 *Hustler,* designated "YB-58" to indicate its prototype development status, rather than production configuration, was manufactured by General Dynamics (GD-Fort Worth) and was the world's first Mach 2 (twice the speed of sound) bomber and first flew in November 1956. With a sleek, "wasp-waisted" (area-ruled) fuselage to reduce aerodynamic drag as it accelerated through the "sound barrier" at Mach 1, four mighty GE J79 afterburning engines slung under a rakish, 60-degree swept back delta wing, it was probably one of the most beautiful airplanes ever built. The airplane, called the "return component," because that is all that would come back from a bombing mission, carried a large pod, containing one hydrogen bomb and extra fuel, slung under the fuselage centerline.

General Dynamics/USAF B-58 "Hustler" Mach 2 supersonic bomber with its warhead pod under its belly—in my opinion the most beautiful airplanes ever built

With a maximum takeoff weight of 163,000 pounds, one forgot that it was only 20% larger (in all three dimensions) than the Convair F-102 from which it got its shape. More fighter than bomber, there was only one pilot—no copilot—with a control stick rather than a yoke, and we other two crew members had separate cockpits in line behind the pilot. We were in separate crew stations, and could not even see each other during fight.

At Edwards we would be conducting the Category II Stability and Control flight tests. I was assistant flight test engineer, working with Lyle Schofield, and the project test pilot was Captain Charlie Bock.

Lyle was a Mormon from Utah, had spent two years in South America as a missionary, and was a quiet and competent engineer. Charlie was tall and husky, with a great sense of humor, and, as I was to find out later on several occasions, an outstanding test pilot.

I was in the second crew station and was responsible for advising Charlie on how much fuel to transfer back and forth among the plane's fuel tanks to control its center of gravity (center of balance). This was critical because of the huge effect that center of gravity had on aircraft performance. If the center of gravity was too far aft, the aircraft would be longitudinally unstable, and if it was too far forward the trim required from the huge elevons at the trailing edge of the wing would create too much aerodynamic drag and reduce the mission range. We had to control the center of balance of 98,000 pounds of fuel to within one foot in a 97-foot long airplane!

I also operated the photo panel and oscillograph to record the flight test data, and kept flight notes. Our crew chief was in the third crew station, behind me, and would keep his eye on the airplane's systems and note any mechanical problems to be investigated before the next flight.

Me in my test engineer's crew station

My test instrumentation panel

Charlie would test the roll, pitch, and yaw rates by pushing the control stick or rudder pedals in various increments. For roll control response, for example, he would bank the airplane to 45 degrees right, suddenly input one quarter left stick to roll the plane 45 degrees left, stabilize, then input one quarter stick right to roll back to 45 degrees right. Then repeat the test for one half stick, three quarters stick, and finally full stick. We would assess the airplane response against the contract requirements established by Wright-Patterson AFB engineers. Charlie would check the airplane's damping by pulsing the controls to see how long it took for the airplane to stop its oscillations. He would also let the plane fly "hands-off" to check its stability by seeing how long it could fly straight and level before it starting rolling off straight and level flight. The airplane was not aerodynamically stable at supersonic speeds so it had automatic systems (roll, pitch, and yaw rate dampers) to help damp out the unwanted—and dangerous—oscillations. Finally he would check the airplane's response to "g" loadings by banking and pulling back on the control stick in a tight turn. Sometimes these were "wind up" turns, in which he would keep pulling back on the stick until the airplane wing stalled. Since the triangular delta wing of the B-58 didn't stall like other wings, the airplane would just start losing altitude, which would terminate this test. We had to conduct these tests at various gross weights, speeds, and altitudes, and with the dampers both on and off.

Once back on the ground the real work began. Charlie would polish his notes, and we engineers would analyze the flight test data, correct it to standard conditions for consistency with other flight test data, plot up the data, and write our reports.

My first flight was exciting: It was my first flight in a jet airplane, my first supersonic flight, and my first flight to Mach 2—twice the speed of sound. Upon landing, the General Dynamics representative gave me a certificate stating that I was the 38th person to fly Mach 2 in the B-58, and a gold "M2" pin to wear proudly on my lapel!

CLUB M2

THIS CERTIFIES THAT

LT. R.K. RANSONE

is a member of the select group of airmen who have flown at a speed greater than Mach Two—twice the speed of sound.

MEMBERSHIP NUMBER 038

Date of qualification 7-18-58

EXECUTIVE DIRECTOR
Bomber Wing

My M2 card, certifying that I was the 38th person to fly at twice the speed of sound in the B-58

Although we flew at almost 1400 mph at almost 60,000 feet altitude, the only time I really felt speed was when we were near the ground. Our takeoff speed at heavy weight was about 246 mph, pretty fast on the ground. When we did tower flybys to calibrate the airspeed and altitude systems we were at about 705 mph and only 50 feet above the runway. The world was really rushing by!

We were conducting these flybys at the nearby Palmdale flight facility, where many contractors had test facilities. After our last flyby, tower radioed: "That's a mighty pretty airplane. Can you make one more flyby for our cameraman?"

Charlie agreed, and we made another high speed pass, and Charlie pulled up into a vertical climb at the end of our pass.

"Thank you, sir," from the tower.

Surprisingly, the interior sounds in the airplane did not change with our airspeed. The only sounds I heard were the sounds of the air conditioning system blowing into my crew station. I barely heard the muffled roar of the jet engines behind me and under the wings.

We had drinking water in a compartment in our canopies, and a long plastic tube for drinking. When we flew subsonic flights at high altitude, the water would be very cold, but when we flew supersonic test flights, the water would become hot from the air friction over the airplane's outer skin.

Over all, my crew station was very comfortable

And it was also great fun!

Guardian Angels

I believe in Guardian Angels. I believe they are among us. And I believe they are us. And I know the name of one of mine: It is Colonel Lane. It was only by the most propitious of coincidences that the closest call I had during my entire career at Edwards, was for a flight that I, fortunately, was not on.

The week before Christmas, in 1958, about a month before Paula and I were to be married, I needed four hours flight time to get my hazard pay for the month. We poor Lieutenants needed every dime we could get, and I needed the money. So when I found that Bob Williams had the back seat open on a B-57 Canberra, I asked for a ride, and he said OK.

Just before lunch, I went to Test Operations and got my parachute and flight helmet, and was waiting for Bob at the dispatch desk, when Colonel Lane, Chief of Flight Test Operations, came by. We chatted for a while, and then he asked me what I was flying.

"The B-57, with Bob Williams," I answered.

He thought for a moment. "That's a pilot proficiency check flight," Colonel Lane said. "Maybe you shouldn't go on that one."

The Canberra was a twin-engine British medium bomber, built by the Martin Aircraft Company, but designed to the British vertical tail criteria of engine out with the remaining engine at idle power for landing, instead of the US criteria of maximum power on the remaining engine for landing wave-off and go-around. This made the Canberra's vertical tail smaller, for less drag, but it created nasty lateral-directional handling characteristics during a landing wave off with

Martin B-57 Canberra Medium Bomber

one engine out when the pilot applied full power on the remaining engine instead of landing.

For some reason, I wasn't really anxious to fly that day, so I said OK, turned in my flight equipment, and left Bob a note that I wasn't going with him. This was undoubtedly *the* most critical decision of my entire life.

I had lunch at the Officers' Club, got a haircut, and returned to the office about 1:00 o'clock.

When I walked in, everyone looked at me like they were seeing a ghost.

"We thought you were on that B-57!" They exclaimed.

"No, I canceled. Why?"

They pointed out the window to the main runway, where an ominous column of black smoke was still rising.

"Bob Williams flipped upside down during a single-engine go-around practice. He was killed."

During my subsequent career at Edwards, I took special precautions on several hazardous test programs, that I believe saved a B-58, an H-41 helicopter, and an NC-130B, and their crews, including me, plus one XC-142A test pilot who would otherwise have been in a fatal crash in Dallas, Texas.

There is a small child alive today because on a trip from Virginia to Texas in 1973 I dashed behind a car backing out of a restaurant parking place and snatched her out of the way. There is another small child alive today whom I ran and grabbed off the back of a station wagon starting to drive off from a gas station in Dallas. The child had climbed through the open back window and was standing on the back bumper and clinging to the back window sill.

I believe in Guardian Angels. I believe they are among us. And I believe they are us. I know because I know the name of one of mine, and because I have been one…

The First Mercury Astronaut

In March of 1959, just before I went to Fort Worth to fly the B-58 heavy weight performance flight tests, Gordon Cooper pulled me aside.

"I just wanted you to know that I have enjoyed working with you, and that I won't be here when you get back."

"Well, I have certainly enjoyed working with you, too, and have learned a lot. Where are you going"

"Does 'Project Mercury' mean anything to you?"

"No."

"Well, I have been selected for a special NASA project. They are going to stick us in little capsules on top of an Atlas Intercontinental Ballistic Missile and fire us into space to fly around the world. Eventually they plan to put some controls on the capsules so that we can steer them a little."

I still remember my immediate reaction, which I did not speak out loud: [*"Are you out of your mind?!?"*]

"Well, that sounds really exciting," I managed to say, figuring that I would never see him again.

I did see him after his flight, however, on several occasions, and we had some interesting conversations.

Edwards was an interesting place, where exciting things were being done by incredible professionals who probably did not realize how unique they were—we were just doing our jobs the best way we knew how, and it was not until later that we realized how special we really were.

Romance

Edwards AFB was a great place to be if you were a young, single woman, but not so great if you were a young bachelor—there were lots of upwardly mobile, single, junior officers, so the pickings were great for a young woman, but pretty slim, romance-wise, for us young junior officers. I had a few dates, but nothing serious. Johnny had gone back to General Dynamics on temporary duty (TDY) to fly on the B-58 performance test program, and had come back to Edwards with a wife and her two year old son.

One Saturday I saw one of our secretaries at the local grocery store who told me that Colonel McBride, the airbase's Deputy Chief Of Staff for Installations (i.e., responsible for base facilities) had his daughter visiting from Texas for the summer. She was going to school at Baylor University, in Waco, Texas. The secretary told me that several of them were meeting her at the commanding general's swimming pool, and that I should come along to meet her. Little did I know at the time how very important this invitation really was.

With nothing better to do that afternoon, I agreed, and wandered up to the pool. The good news was that Paula McBride was intelligent, refined, interesting, her background was similar to mine, she laughed at my jokes … and she was also *gorgeous*! The bad news was that every other single, junior officer on the base thought so too! The bad news was that she was engaged to someone back in Waco, but—more good news—she was dating during the summer, which suited me just fine. We could go out on dates and I was not likely to get married because she was engaged already. The only bad news was that she didn't want to kiss good night—or anything—while she was engaged.

We shared a lot of common interests, and went to musical performances in Los Angeles (about an hour and a half's drive from Edwards). Paula also like to play bridge. Sometimes I would see her father at happy hour at the Officers' Club, and he would invite me to their quarters to play bridge. He was an excellent bridge player, and remembered every card that had been played. Her mother played well, also, but I played rather non-traditionally. For example, if I had a queen and only one low trump, it meant that I had no protection for my queen and he could call it in anytime he wanted by leading his ace and king. In this case, I would throw my queen on the first round instead of my low trump. He would assume that I was out of trumps, and play away, expressing great surprise—and also great dismay—when I trumped in on a crucial trick and took the rubber. As a result, he always sat to my left at the table. After he caught on to my gambit, he was still confused because he didn't know whether I was out of trumps or not.

Great fun.

Then, one day Paula told me that we couldn't date that weekend because her boy friend was visiting from Texas. I saw her mother and father at the base theater. This was unusual because they didn't normally go to movies, so I figured something was up.

Indeed it was! The engagement was off, and Paula decided not to return to Baylor in the fall of 1958.

That first kiss was ***wonderful***!

We dated a lot during the fall of 1958, but she was also dating other junior officers. One day a junior officer named Bob Nagle, who worked in her dad's office, told me that "...the Colonel is really pushing me with Paula!" Paula's dad was "pushing" Bob Nagle for Paula? That was the best news that I had had all week. If there was one thing I had learned about young women, it was that to have a parent "push" a particular mate was the *Kiss of Death*! So I gloated, while Bob basked in a state of blissful ignorance.

And so it was.

A side bit on Bob Nagle: He was a jock, and a New Yorker, and he was "pushy," (but, as Mark Twain said about politicians and idiots: 'I repeat myself'). Actually, Bob was OK, in his way, but he had a habit of helping himself to any "goodies" that we might have sitting out in our rooms. One day I saw a box of rubber chocolates in a novelty store. They looked and smelled like the real thing, but they were foam rubber. I set them out on my chest of drawers, and, in due course, Bob wandered in, spied the chocolates, and helped himself. About halfway through his third one, he remarked "These chocolates taste funny!" "That's because they are *RUBBER*!,"I exclaimed, with great glee.

Back to Paula and me: in December, Paula and I decided to take Latin Dance lessons in town. We saw an ad in the paper for classes, and enrolled for ten lessons. At the first lesson, however, we discovered that we were the only couple to sign up for the classes, but the instructor had expected a dozen or so couples. After the second lesson, we agreed with the instructor that it was not profitable for her to give us private lessons for the price of a class, so she refunded our money and we stopped going.

The instructor, by the way, was quite a *looker*. Very pretty, in a bright red dress with a wide neckline and a high side slit. I enjoyed this lesson a lot better than Paula did.

That night we became engaged to be married —Paula and me, not the instructor and me! After I left Paula at her quarters, I sat out under the stars on the golf course and thought of my future with this wonderful woman. I was very excited, and wrote Mother all about it. I deliberately wrote her that very night so that my happiness and excitement would come through in my letter, and Mother said that it really did. She kept the letter for a long time. I looked for it after she died in order to preserve it, but I couldn't find it. I like to say that the best place to meet your future wife is a swimming pool, because that way you can see what you're getting. Paula saw what she was getting, too, but surprisingly, she married me anyway.

And that night in November 1958 I went to bed happy, with a big, silly grin on my face.

Wedding Day, January 30, 1959

We Start B-58 Heavy Weight Performance Flight Testing

The B-58 was ahead of its time—perhaps it even ventured too far "towards the unknown" in many respects, because it had twice the accident rate of the notorious Lockheed F-104A *Starfighter*. During the year that we did the Category II performance tests out of GD's plant next to Carswell AFB in Fort Worth, of the 30 airplanes flying, six were lost along with half of their crews. This was about the same kill rate as an infantryman on his third tour of duty during the Vietnam War. I was never afraid, once we got into the air and I was busy with my test responsibilities, but I frequently had butterflies before a flight, and they were much worse if we had been down for a long period. I always said a short, silent prayer before each flight, asking that all those flying would return safely, and after a safe flight, I always said "thank you."

And it didn't help that, every time a B-58 went in, Mother, who lived in Fort Worth and heard the news on the television, would call to see how I was. It didn't matter that I had just telephoned her. She always had to reassure herself of my safety.

And it also didn't help matters when our test pilot quit. "There are too many things going on in that airplane that I don't know about—I can't even see the wings!" This was immediately after one airplane ran off the end of the runway and exploded, and another just blew up on the ramp during preflight. If he was really that scared, I was glad that he quit, because I thought that if we got into a bad situation he might not be able to think clearly and logically, and could in fact endanger the airplane and our crew. I also admired him for having the guts to admit it.

Then Charlie Bock, with whom I had flown on the stability and control tests at Edwards, became our test pilot. Good move. A superb test pilot.

He demonstrated how good a pilot he was during aerial refueling. To extend our flight times we frequently refueled behind a Strategic Air Command aerial tanker. This is a very tricky maneuver: The two airplanes are flying within about 20 feet of each other at 350 knots or so. Airplanes normally climb and descend a little as they fly. A good tanker pilot will let these small variations in altitude take care of themselves, and the planes will slowly descend and climb together. If the tanker pilot tries to compensate and pulls back on the stick to arrest a slight descent, it lowers the tail and the tanker climbs. The receiving airplane behind has to lower its nose to clear the tanker's lowered tail, which causes it to descend. The result is an automatic disconnect as the tanker climbs and the receiver descends. The tanker and receiver pilots had to work as a team. The problem was exacerbated during refueling because the tanker was losing weight and trying to accelerate or climb, and our airplane was gaining weight and trying to descend or decelerate.

Me in my Flight Test Crew Station

This was complicated even more with the B-58 because of the large amount of fuel being transferred, up to 90,000 pounds or so, over half the weight of the entire airplane. The tanker was flying close to its maximum airspeed and we were flying close to our slowest speed because of the different airplane designs. To stay connected, Charlie had to slowly increase engine power as we gained weight, up to the point at which he had all four engines at maximum power without lighting the afterburners. As we gained more fuel, he had to light two of the afterburners. This extra fuel injected into the engine exhaust provided enormous extra thrust that occurred suddenly, so Charlie had to immediately reduce the power on the two non-afterburning engines. He did this with the hands of a surgeon! On one particularly good day we were behind an excellent tanker pilot. We took on 90,000 pounds of fuel and never got a disconnect. Charlie complemented the tanker pilot, and the tanker pilot replied to Charlie "That's a mighty fine bit of flying back there, yourself." And indeed it was.

Surprisingly, I was not overly concerned with my test flying. We were all professionals, properly trained, knew our airplane's capabilities and limitations, planned our flights carefully, and knew what we were doing. For example, I assessed things as I observed them to be, not as I hoped they were. I always tried to figure out what could go wrong and be prepared for it. Flight testing was not at all like the Hollywood stories where Clark Gable climbs into a new airplane and "takes her up to see if she flies!" Thousands of hours of calculations, computer analyses, and wind tunnel studies had mostly defined what the planes could, should, couldn't, and shouldn't do. We were trained in all airplane systems, understood everything that was going on in that very complicated airplane, new our jobs thoroughly, carefully planned our tests, and followed the test cards exactly.

A good example of our AFFTC's professionalism was demonstrated one day when our crew chief, Sgt. Dick "Jenkins" (sorry, Dick, but I don't remember your real last name), watched the General Dynamic crews servicing and pre-flighting our airplane. As they were checking the hydraulic systems they found a leak in one of the systems. The airplane had two, completely independent hydraulic systems. Since the airplane could not be flown without its hydraulic systems operating the flight controls and other systems, it had two independent hydraulic systems for safety. In the event one failed, the other would get us home safely.

At least they were supposed to be independent: on this occasion, however, when the mechanics fixed the leak and bled the system, hydraulic fluid came out of the other system. Dick told them that that was not supposed to happen. They said it was OK, since they were bleeding the system, but Dick persisted. They investigated and found that a critical part of the two systems had been installed backwards, effectively connecting the two systems into one system, so that if we had lost either hydraulic system in flight, we would have lost the airplane.

Good work, Dick.

"Hey, Rob. Got any extra fuel in your flight station?"

Our Edwards flight test crew was composed of our project flight test pilot Captain E.E. Bradley, (later replaced by Captain Charlie Bock), project engineer Carl D. Simmons, me as flight test engineer, and our crew chief Staff Sergeant Dick Jenkins. We set up our office in a space provided by GD who also provided an engineering aid and secretarial services. Carl, E.E., and I planned our flights in accordance with the detailed flight test plan that Carl had written before leaving Edwards. GD serviced our airplane under Dick's close supervision, and E.E., Dick, and I flew the test missions. Carl had a bad back, and believed that he should not risk having to eject in the event of an emergency. E.E. followed the test card that we had prepared and that GD had approved (for flight safety). I operated the flight test instrumentation to record the flight test data, took notes, and advised E.E. on the next test condition and fuel transfers to maintain the correct airplane center of gravity. Dick monitored the airplane systems and made notes on anything that needed attention after the flight.

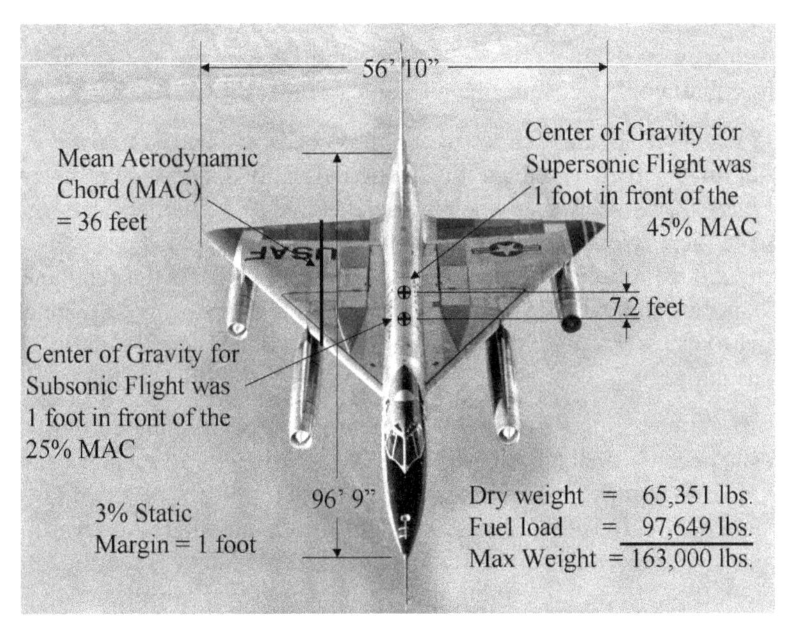

B-58 Dimensions, Showing Center of Gravity Limits, Adjusted by Transferring Fuel

Airplane center of gravity (center of balance) is very critical for a delta-winged airplane with no horizontal tail, so I monitored our c.g. and advised E.E. on fuel transfers between fuel tanks needed to maintain or set up the correct c.g. Later airplanes had automated c.g. control for which the pilot could set the desired c.g. and the system would maintain it, but in 1959 we had to do this manually. The biggest shift was when we accelerated from subsonic to supersonic flight, or decelerated from supersonic to subsonic flight. We transferred fuel in order to maintain a 3% static margin for longitudinal stability and low aerodynamic drag. That's balancing 97,649 pounds of fuel to within one foot!

This was necessary because of the way the center of lift on the wing changed between subsonic and supersonic flight. At subsonic airspeeds the center of aerodynamic lift on the airplane's wings was located about 25% back from the wing leading edge. At supersonic airspeeds, however, the aerodynamic lift was located about 45% back from the wing leading edge. If the airplane's center of gravity was in front of the center of lift, the airplane would be stable longitudinally, and balanced by the wing trailing edge elevons, but if the center of gravity was too far in front of the center of lift, the elevons would have to be deflected too much, and create too much drag that would decrease the airplane's range.

So when we accelerated to supersonic speeds, we had to transfer fuel aft, 7.2 feet, to reduce the elevon deflections, and then transfer it back forward 7.2 feet when we decelerated to subsonic speeds to make the airplane longitudinally stable. I had a small hand calculator (not electronic in 1959) on which I lined up the fuel tank quantities and calculated the c.g. I could then advise E.E. how much fuel to transfer, and which way, in order to control our 3% static margin.

In-flight refueling was a major element in our performance flight tests, because the complicated B-58 was so difficult to maintain and we got so few flights a month. Once we got in the air we hated to come down. The Air Force's Strategic Air Command (SAC) was very cooperative and supported us out of Carswell AFB for almost every flight to provide training for its tanker crews. Unfortunately, the B-58's fuel gauges were not accurate enough to determine our test weights if we refueled in the air during a test flight.

I studied a plan of the fuel lines, and determined that two fuel flow meters, strategically placed, would allow us to meter the fuel as we received it from the tanker. We requested General Dynamics to install these flow meters, which they did. The only catch was that we did not know how much fuel went into the pod tanks until we transferred it all up into the airplane itself. No problem—we established a routine of emptying the pod tanks into the airplane, through the fuel flow meters, before landing.

Except on one flight, after which, when I was studying the post flight fuel counter data, I noticed that 5000 pounds of fuel had left the pod, but never arrived in the airplane. After marveling at this for a few minutes, I called the crew chief.

“I’m missing 5000 pounds of JP-4 that disappeared between the pod and the airplane.”

“No it didn’t—I just found it,” he replied. “It’s sloshing around in the belly of the airplane!”

"...And Wash Our Windshield, Please..." At 25,000 Feet

Our supersonic, heavy weight performance flight tests of the B-58 out of General Dynamics' Fort Worth plant, sharing runways with Carswell AFB, were limited in 1959. We had a special supersonic corridor, approved by the FAA, along which we wouldn't annoy too many people with our very loud and sudden sonic booms. We would fly north to Liberal, Kansas, turn south, and accelerate to Mach 2. We had only a few minutes of test time before we had to turn around at the Gulf of Mexico, since we were not cleared to fly over water in these prototype airplanes.

To give us more supersonic flight test time, I conceived a plan to fly to Edwards, conducting subsonic performance tests on the way, refuel from a SAC KC-135 tanker over Edwards, then fly supersonically back to Fort Worth. If we had a problem refueling, we could land at Edwards because another B-58 was being tested there, and there were ground support equipment and personnel.

All went well until we tried to refuel over Edwards at 25,000 feet. The refueling receptacle was in the nose, directly in front of the pilot. When Charlie Bock engaged the toggles to clamp down on the KC-135's refueling probe and lock it into place for the fuel transfer, a white fog suddenly erupted from the receptacle, and immediately all four forward windshield panels were coated with hydraulic fluid. The tanker was visible only through the upper canopy windows. Charlie was essentially flying blind.

Charlie disconnected from the tanker and closed the refuel receptacle. With no windshield wipers (on a Mach 2 airplane?) we didn't know how long it would take for the wind blast to clear the windshield. Supersonic airflow might do it, but if anyone wanted to go supersonic with a known hydraulic leak, he wasn't on *our* airplane!

Charlie said, over the intercom: "I can't see well enough to land the airplane!"

After some thought, I suggested: "Charlie, maybe the tanker can wash our windshield with a cupful of JP-4."

Charlie answered: "I've been thinking the same thing."

So he asked the tanker to wash our windshield. Carl Simmons was on the tanker, and he said the tanker crew was stunned.

"What if we get fuel into the engine inlets—we'll all blow up!" Carl told them that the guys back in the B-58 hadn't just fallen off a turnip truck, and knew what they were doing.

"If they didn't think it safe, they wouldn't have asked."

The operator uses the black vanes on the probe to "fly" the boom into the refueling receptacle in the B-58/s nose

The tanker pilot finally agreed, and the boom operator talked Charlie back under the boom, and, with one small squirt, cleaned our windshield slick as a whistle! We then landed safely at Edwards, where I lived with the McBrides for two days in my flight suit.

From Edwards, we did a maximum performance takeoff and accelerated immediately to Mach 2 towards Fort Worth. We passed Albuquerque, started decelerating over west Texas, and arrived over Fort Worth only 45 minutes after brake release at Edwards. We got all of our Mach 2.0, 1.8, and 1.6 cruise performance done in one stretch. Since we had blown off the fuel jettison vent cover, we had to cover up the hole in the aft section of the fuselage, but we didn't have another fuel jettison cap with us. So the crew chief just covered the hole with duct tape. OK, so it was stainless steel duct tape—bit it was still duct tape, and it was still on after our supersonic cruise back to Fort Worth.

We landed at Fort Worth just as another B-58 was taking off. Moments later it disintegrated over southern Oklahoma during a special Mach 2 engine shutdown test at 36,000 feet, killing its two-man crew.

Their test mission was to prove that the B-58 could lose an outboard engine at Mach 2 and 36,000 feet altitude (the lowest altitude at which the aircraft was designed to fly Mach 2, because of the highest aerodynamic air pressure on the structural airframe). Because it was a hazardous, structural flight, the plane had a minimum crew of only two, the pilot and the flight test engineer.

All went well, they established the test condition, and shutoff the fuel flow to the number four engine (the starboard most engine). It is not clear what happened next, but most assessments indicate that the number three-engine, inboard on the same side, also shut down. At any rate, a photo from the ground shows the plane's contrail tracking straight, then veering slightly as the number four engine was shutdown, then veering violently more, an explosion's white cloud, and pieces of the airplane plummeting to earth.

The high vertical tail causes rolling moments when the rudder is deflected to counter yaw

Subsequent assessment of the recovered parts of the airplane indicated that the tail had twisted off. Since the vertical tail was high above the fuselage, any large forces on the tail would also input a high rolling moment, or twist, to the aft fuselage. The air loads at Mach 2 and 36,000 feet were about 1,200 pound per square foot, and the limit sideslip angle was only 1 ¾ degrees. This doesn't seem like much, but I've been there, and I was literally hanging on my shoulder harness from the high side loads. Very uncomfortable!

The aft fuselage looked like some child had taken his toy airplane, and in a fit of pique, had twisted off the tail. The nose section, about six feet long, was broken in three pieces in front of the cockpit. The crew was ejected into a 1,400 mph air stream, 1,200 pounds per square foot pressure, and about 600 degrees Fahrenheit caused by the air compression on their bodies. The pieces of their orange fight suits were bleached pure white.

After many investigations, calculations, and meetings, the B-58 System Program Office at Wright-Patterson AFB restricted the aircraft to Mach 2 at no lower than 42,000 feet altitude, where the dynamic air loads were less severe.

Cruising serenely along at 1,400 miles per hour, cozy, comfortable, and busy in our aluminum cocoon, we were not always aware of the terrible forces lurking only a foot away, just on the other side of our tiny, thin Plexiglas windows, just waiting to destroy our airplane and tear us to pieces if we made a mistake.

Fortunately, we didn't make any mistakes.

A quick lecture on supersonic aerodynamics

When anything makes a sound, what is actually happening is that the air pressure is compressed and propagates outwards, much like the bow wave as a boat travels through the water. This pressure waves travels at—you guessed it—the speed of sound, which is about 742 MPH on a 59°F day at sea level. Your brain interprets those pressure waves as "sound." In case you are wondering about the age-old riddle: "if a tree falls in the forest with no one around, does it make a sound?" The answer is "no." It makes an air pressure wave, but with no one (including animals) to interpret the pressure waves, there is no "sound."

When an airplane flies faster than the speed of sound (called "compressible flow" in aerodynamic terms because the air cannot propagate the pressure wave out of the way fast enough) it creates a shock wave. You can make this sound yourself by cracking a whip—the tip of the whip is faster that the speed of sound, and the "crack" is a shock wave.

B-58 shown at Mach 2 (twice the speed of sound) with simulated shock waves that "boomed" the ground when we flew supersonically

The angle of the shockwave is directly related to how fast the airplane is going. The ratio of the airplane's speed to the speed of sound, is called Mach number, named for Ernst Mach, an Austrian, in about 1870. Mach 1 is the speed of sound. The equation for the shockwave angle, measured from the longitudinal axis of the airplane, if you are still interested, is:

$$\textbf{Shockwave angle} = \sin^{-1}(1/\text{Mach Number})$$

So at Mach 1, the wave angle is 90 degrees (i.e. vertical), and at Mach 2 the wave angle is 30 degrees, as shone in the picture.

The position of the spikes protruding from the fronts of the engines is critical: The spikes moved in and out automatically in order to keep the shockwaves just touching the lip of the engine inlets. If the spike is too far out, the shockwaves miss the inlets, and air spills around the engine and the engine can't get enough air. If the spike is too far in, the engine inlet "swallows" the shockwave, a normal (vertical) shock wave occurs in the inlet and the engine gets what is called an "un-start," and flames out.

When the sun was just right, I could actually see the shockwaves from the engine spikes touching the engine inlets, but I could never get a picture of them.

What fun!

Two last comments about supersonic aerodynamics:

First, the aerodynamic center (i.e., center of lift) on the wing is usually located at the 25% mean aerodynamic chord (MAC) (i.e., 25% of the average distance from the wing leading edge to the trailing edge) during subsonic flight, but shifts back to about the 45% MAC for supersonic flight. This is why we had to transfer fuel back and forth during supersonic flights.

Second, the faster the Mach number, the less effective the vertical tail and rudder become (in coefficient form: Cn_b). When Mel Apt disintegrated in the Bell X-2 at more than twice the speed of sound, he had performed one of the most perfect flights yet in the airplane. As a result, he found himself too far from the Edwards lakebed, and pulled back on the control stick to aim toward the base. At that Mach number, his Cn_b was already very low, and when he pulled too many gs, this exacerbated the situation, Cn_b went to zero, and it doesn't matter how much aerodynamic pressure he had, zero times anything is still zero: the X-2 tumbled and disintegrated, killing Mel Apt.

Now a short lecture on the area-rule, and how it reduces aerodynamic Mach wave drag. The most impressive demonstration of its effect is with the F-102 delta wing fighter. The YF-102 had a straight-sided fuselage, and was capable of only M 0.98. When the fuselage was "pinched-in" in the middle in accordance with the "area rule" the top speed increased to M 1.22.

At high-subsonic flight speeds, supersonic airflow can develop in areas where the flow accelerates around the aircraft body and wings. The speed at which this occurs varies from aircraft to aircraft, and is known as the critical Mach number. The resulting shock waves formed at these points of supersonic flow can bleed away a considerable amount of power, which is experienced by the aircraft as a sudden and very powerful form of drag, called wave drag.

To reduce the number and power of these shock waves, an aerodynamic shape should change in cross sectional area as smoothly as possible. This leads to a "perfect" aerodynamic shape known as the Sears–Haack body, roughly shaped like a cigar but pointed at both ends.

The area rule says that an airplane designed with the same cross-sectional area *distribution* in the longitudinal direction as the Sears-Haack body generates the same wave drag as this body, largely independent of the actual shape. Since aircraft are configured so that large cross sections, like wings, are positioned at the widest area of the equivalent Sears-Haack body, and the cockpit, tail plane, intakes and other "bumps" are spread out along the fuselage, then the rest of the fuselage along these "bumps" must be correspondingly thinned.

The area rule was discovered by Otto Frenzl when comparing a swept wing with a w-wing with extreme high wave drag while working on a transonic wind tunnel at Junkers works in Germany between 1943 and 1945. He wrote a description on 17 December 1943, with the title *Arrangement of Displacement Bodies in High-Speed Flight;* this was used in a patent filed in 1944.

Richard T. Whitcomb, after whom the rule is named, independently discovered this rule in 1952, while working at the NACA, Langley Research Center. Whitcomb had a "Eureka" moment. The reason for the high drag was that the "pipes" of air were interfering with each other in three dimensions. One could not simply consider the air flowing over a 2D cross-section of the aircraft as others could in the past; now they also had to consider the air to the "sides" of the aircraft which would also interact with these stream pipes. Whitcomb realized that the Sears-Haack shaping had to apply to the aircraft as a whole, rather than just to the fuselage. That meant that the extra cross-sectional area of the wings and tail had to be accounted for in the overall shaping, and that the fuselage should actually be narrowed where they meet to more closely match the ideal.

The area rule increased the maximum speed of the YF-102 from M 0.98 to M 1.22, and was critical in allowing the B-58 to fly supersonically.

The area rule was most effective in the transonic region from about M 0.95 to M 1.3. In fact, the B-58 could fly with less power at M 1.2 than at M 1.0.

I have often thought about how the Area Rule works, and have a theory:

For subsonic aeronautics (i.e. incompressible flow) the air pressure on the aircraft changes with airspeed, in accordance with Bernoulli, but the density does not. For supersonic aeronautics (i.e., Compressible flow) the air density changes as well as the air pressure. The problem occurs because the air density ratio is changing faster that the air pressure ratio.

I suddenly remembered that during my senior year at A&M, as a part time job with the aero department, I was drawing the construction blueprints for its new supersonic wind tunnel. I was designing and drawing the tunnel, IAW the aerodynamics professor's instructions, but he said that he would draw up the throat, because of the vary precise shape it had to have in order to avoid a normal shock and choking.

As you may know, as the throat is reduced in size, the air speeds up until it reaches M 1, at which a normal shock occurs and the tunnel chokes--UNLESS the throat is expanded along a precise equation, as which the air speed continues to accelerate supersonically.

Remembering this, I rationalized that in flight, if the air is accelerating over the aircraft body, at some point it must be ripe for a normal shock, choking, and increased drag. Now, what if the aircraft shape were allowed to change along a precise formula that would allow the airflow to continue to accelerate (similar to the expanding throat of the supersonic wind tunnel)? This means the airplane shape must not have any sudden changes in cross-sectional shape, especially a sudden increase in shape, similar to shaping the throat cross-section in the supersonic wind tunnel.

This would mean that the aircraft fuselage should be reduced in cross section when other aircraft components are added, such as engines on the B-58 and vertical tails on the F/A-18 E/F. In other words, the area rule.

Some One-Dimensional Isentropic Compressible-Flow Functions data are shown in the following figure. Note that above M 1.0 both pressure ratio and density ratios are decreasing with Mach Number, but that the density ratio divided by pressure ratio is increasing, and following the increasing Mach Number almost directly.

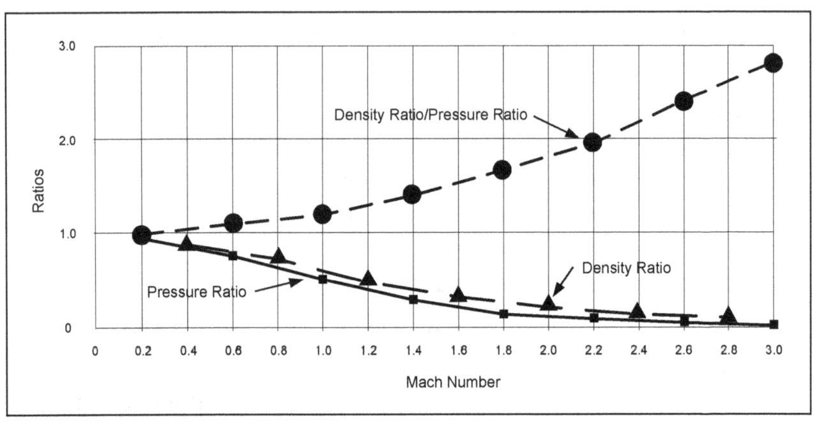

Note that the pressure and density ratios are decreasing with Mach number, but the density ratio divided by the pressure ratio is following the Mach number exactly above M 1.2

I suspect that the reason the area rule works, is based upon the same Mach wave phenomenon that required a precise expansion ratio for the throat of a supersonic wind tunnel in order to prevent a normal shock wave and throat choking. The cross-sectional areas must vary in order to accommodate the changing density for compressible flow.

Now you know more about compressible flow than you really wanted to know.

So why did three pros almost jump out of a perfectly good airplane?

Some amusing things happened in the B-58 that entertained us during our long "boring" flights. For example, on one flight, as we were returning to base, Charlie Bock mentioned over the intercom that he had just lost the aft fuel booster pumps. The fuse panel was in my flight test engineer's station, so I announced that I would check. Sure enough, the aft booster pump fuse was blown. I looked around for a spare.

"Charlie, I think I've found a spare fuze the right size. There was something written next to it that has been erased. I think it's a spare. I'll pull it, and if anything happens, you holler."

Charlie said "OK."

I pulled the fuse.

Immediately there was loud, excited hollering in our earphones. Then silence.

I was probably the calmest of all of us because I was the only one who knew exactly when I pulled the fuse, and nothing obvious had happened, no lurch of the airplane or change in aircraft engine or systems sounds. But we waited expectantly for Charlie to say something.

Dick, in the third crew station asked, tentatively:

"Uh. . .that wasn't you, was it Charlie?"

Charlie said "No. It scared the Hell out of me!"

The fuse blew again, but we still had alternate fuel pumps, so there was no immediate problem. We landed without further incident, but we never found out who had keyed his mike and yelled at that exact moment, or why.

Great Data From A Lot Of Airplanes

Not all of the thrills of flight test come in the air, sometimes they come from some clever discovery during data analysis. Non-flight test engineers won't believe this—just take my word for it. For example, to determine the altitude at which to top out our Mach 2 check climb plots that I was preparing, I had to know the altitude at which the B-58A started its cruise climb mode. "Cruise climb" occurred when the airplane reached its absolute ceiling for its weight, but was burning fuel so fast that its decreasing weight would allow it to gradually climb to higher altitudes. I was reducing the climb and acceleration data, and Carl was reducing the cruise performance data.

B-58 In Flight With Weapons/Fuel Pod

We had determined that the best Mach 2 range was at maximum afterburner power, letting the airplane climb at constant Mach Number as it burned fuel. Don't scoff—fuel was galloping through the four engines at about 1,500 pounds per minute! So I needed the Mach 2 cruise performance data in order to finish my climb to ceiling plots. This was going to take a while for Carl to complete because of the complex calculations needed to correct the data to standard conditions. Furthermore, these corrections depended upon analysis of our own flight test data, that would not be ready for several weeks.

In order to expedite matters, I collected all of the data that I could find from other GD B-58s flying at Mach 2 at maximum power and cruise climb. All I really needed was their gross weights, altitudes, and the ambient air temperature. I collected data snippets from about a dozen flights, and plotted them as altitude versus temperature for lines of constant standardized weight (w/δ, for you technical types). The lines for individual airplanes showed much steeper slopes and variations than the aggregate trend, so I used the aggregate trend data as being more representative of the B-58 fleet. This was before the days of powerful little computers that come in cereal boxes, so I had to draw slopes and temperature corrections by hand. (Yes, I could have done it on the latest IBM 704 computer, for which a programmer would have written a long program—in "octal"—and I would still be waiting for the answers).

Guess what? the data came out fine. My cruise ceilings were within eight tenths of one percent of the "properly" corrected cruise climb data that we got out of the computers six months later.

What fun!

Advanced Schooling Courtesy of the U.S. Air Force

There were wonderful opportunities for advanced education at Edwards.

The first course I took shortly after I got to Edwards was a short course on Space Technology, taught at the new Thompson Ramo-Wooldridge Inc., aerospace company in Pasadena. Simon Ramo and Dean Wooldridge formed the Ramo-Wooldridge Corporation and merged with Thompson Products in 1958 and were very active in the early space and missile programs. The name was shortened to TRW Inc. in 1965.

Several of us drove down once a month to Pasadena to attend these lectures that discussed the solar system, astronomy, and space mechanics (how celestial bodies moved). This was very interesting but I never used that knowledge to any useful purpose.

I also signed up for an Advanced Mathematics course taught on base by a professor from the University of Southern California.(USC). I have always thought that there are no poor students, only bad teachers, and this professor confirmed that suspicion. Each week we had to study a very difficult book on non-linear differential equations, work the problems, and then he would swoop in once a week and answer our questions! I had no clue what was going on or how to do the equations. I had had differential and integral calculus and differential equations at A&M, but mathematics in which you had to change an unsolvable equation into a different equation that you could solve, and then transfer the solution back to the first equation was way beyond me.

I was absolutely snowed, so when USC notified me that my over all grade point average from A&M was not good enough for its graduate program (I had a B average in my technical courses, but a C average overall because of the difficulties I had when my father died in 1953) I was not disappointed to drop out of that miserable course. Obviously I got no useful value from that course either.

The Air Force assigned me to an on-base Management Development course that included working with others—so-called "sensitivity training"–that I thoroughly enjoyed and that helped me not only at Edwards, but later throughout my career. I was also sent to an Executive Development course at Texas A&M that helped me, also. I took a 16mm film on the flight test center that I showed to the aero students that they really enjoyed.

I learned management techniques from these two short courses that I applied in my business relationships and also in my personal life. I had to be especially cautious applying these management principles at home, however, or Paula would accuse me "Don't *manage* me!" They proved most useful later at American Airlines dealing with hostile citizens groups and especially when I became a business consultant and had to motivate my clients' people over whom I had no direct control, and many of whom didn't want me there in the first place.

The Height of the Cold War

The early 1960's was the height of the Cold War with the Russian Block countries. Long range, U.S. B-52 bombers were on constant alert, and there were actually many of them constantly in the air, armed with nuclear weapons, to be diverted to their Russian targets if the Russians should attack. We also had nuclear-armed intercontinental ballistic missiles with Russian target addresses already programmed into their inertial guidance systems. The Russians had put "Sputnik" into orbit, and on clear desert nights we could see its faint reflection as it passed overhead—a grim wake up call that we were no longer out of reach and no longer invulnerable.

About ten miles from us was a small, desert town named Boron, famous for its nearby borax mines and "Twenty Mule Team Borax," a cleaning product. These mines were very deep, and would be secure in the event of a Russian nuclear attack. The Air Force had stocked the mines with food, water, and other supplies, and we were alerted that if the base sirens sounded, we were to go immediately to the mines because a nuclear attack was imminent.

On the hills around large cities and military bases we could see ground-to-air missile batteries, poised and ready for defense. This was the time when many civilian families dug bomb shelters in their back yards, and companies offered packaged survival supplies of food, water, flashlights, battery-powered radios, and other supplies.

These were very scary times, but eventually everything settled down. I think that when President Kennedy called Russian Premier Kruschev's bluff during the Cuban Missile Crisis, was the turning point. Kennedy had sent high-flying Lockheed U-2s over Cuba, invulnerable to attack because of their high altitude, and the photos they brought back of the Russian missiles provided conclusive proof of the missiles' presence near our border. The photos printed in the newspapers were impressive due to their clarity, but the actual classified photos were even more incredible because they reportedly showed readable headlines on the newspapers that Cubans were reading!

Just before coming to Edwards, Paula's Dad had built the top-secret base in northern Norway, where the U-2s landed after over flying Russia.

And This Baby Will Accelerate to 197 mph And Stop in 2 ½ Miles

At Edwards, I was still working with the B-58.

Another really sporting thing about the B-58 was that if you lost an engine on takeoff at, say, about 150 knots, you weren't going fast enough to take off on the remaining three engines, but you were going too fast to stop if the drag chute didn't work. The drag chute was a small parachute attached to the rear of the airplane, deployed on landing to help slow the plane. But no matter how much runway you had left, the brakes could only absorb 18 million foot-pounds of energy each. Period. That's enough energy to instantly boil 21 gallons of water from each brake, and there were eight brakes! But even these brakes were not powerful enough to stop fully loaded airplane without the drag chute's help. To put that amount of energy absorbing braking capability in perspective: one B-58 brake could stop an automobile traveling at the speed of sound—742 miles per hour! But this limitation for the B-58 left a significant takeoff "dead man's zone" where, if you lost an engine and the drag chute failed, you couldn't takeoff and you couldn't stop. You would run off the end of the runway, which would spoil your whole day!

B-58 landing with drag chute braking

Takeoff speed at maximum weight was 214 knots (about 246 mph) —just below the 217-knot speed at which the quarter inch-thick, 22-inch diameter tires, inflated to 275 psi, were designed to start flying apart at 6000 rpm. These tires and wheels had to be small in order to retract into the thin supersonic wings in which there was not room for larger wheels and more powerful brakes. It was a "Catch 22."

The small wheels and tires were continuing problems, and two accidents were directly related to them. On the first, Captain Ken Lewis was flying out of General Dynamics when a tire blew during takeoff. The wheel started disintegrating, Ken was advised by the control tower, and aborted the takeoff, but ran off the end of the runway, the landing gear collapsed, and the airplane exploded. Ken escaped by climbing over the cockpit windscreen, running long the nose and dropping to the ground. The second and third crew members were not so lucky—they ejected, but landed back in the fire and were killed.

On another instance, Lt. Col. Joe Cotton, took off from Carswell AFB on a night flight. The control tower noticed that, where as there should have been four fires blazing from the afterburning jet engines, there were, in fact five fires. A tire had blown, the wheel had disintegrated, and blown holes in the wing and fuel was dripping down on the burning wheel. Joe continued the takeoff and accelerated, at which time the wind blew out the fire.

A quick conference to decide what to do. The decision was made to fly to Edwards AFB to land on the 15,000-foot main runway, with the 15-mile dry lakebed overrun. SAC scrambled aerial tankers to rendezvous with the B-58 because the wing fuel tank had been punctured and the fuel was leaking very badly. Ken flew all night, and refueled eight times on his way to Edwards AFB. Remember that, even though he had an autopilot, the B-58 had only one pilot station. It was a very grueling flight.

Joe arrived at Edwards early the next morning, just at sunrise, and one of the Edwards B-58 pilots flew chase to look him over. We had foamed the main runway with patches of foam periodically along the main runway, and concentrated the foam where he was expected to stop.

He made the landing safely, and saved the airplane.

Back to our program: The B-58 refused takeoff speed had never been demonstrated to the maximum brake limits, so, in November 1960, Fitz Fulton and I got to do it!

Fitz was another great test pilot. For example, when he flew the first Air Force flight tests of the B-58 in Fort Worth, he demonstrated his cool professionalism: On one flight, with Johnny Armstrong in the back seat, they lost an engine at Mach 2. Remember what happened to the airplane over Oklahoma a few years later. In his test report, he described his reaction: "My first reaction," he wrote, "was to hit the rudder pedal to correct the violent yaw, but I realized, instantly, that the airplane had reached its maximum yaw and was coming back. I was concerned that if I pushed the rudder that I might get in-phase with the oscillations, increase the yaw, exceed the rudder hinge moment limits, and we would lose the airplane. So I just hung on, letting the oscillations damp out naturally, and then retarded the opposite engine throttle. After we had stabilized, I reduced power on the remaining two engines, and we decelerated to safe, subsonic airspeeds."

Fitz was also a wonderful family man. I have seen him interrupt a flight planning meeting to call home and ask his wife if she wanted to go to the movie that night. His test airplane had his daughter's name painted on the nose.

To answer your obvious question: Yes, he survived more than 20 years of experimental flight testing and retired from the Air Force. He went on to fly more than 240 different airplanes, almost every exciting plane the Air Force had, including the XB-70, the SR-71, and years later the Anglo-French Concorde airliner. He flew NASA's Boeing 747 that ferried the Space Shuttle from Edwards AFB to Cape Kennedy after Space Shuttle landings at Edwards. During his long career, he logged more than 16,700 hours of flight time.

Back to our B-58 refused takeoff tests: Our mission: Demonstrate a refused takeoff, at which all four engines are shutdown at the three-engine "go speed," and stop without using the drag chute, generating the design brake limit of 18 million foot pounds of energy on each of the eight brakes. There are proven engineering equations that define energy. When an airplane (or your automobile) is rushing down the runway (or highway) it has a certain amount of kinetic energy that is a function of its weight and the square of its speed. When you stop, all of that energy has to be absorbed somehow. The "somehow" is into the brakes as heat. The faster you go, the more heat. Automobile disk brakes are much better than the old drum brakes because they cool better, but even now, after several high-speed stops, the brakes may "fade" due to the heat buildup. Taken to the extreme, eventually the brakes will get too hot, lose all their friction, and can no longer stop the vehicle. This is what we had to determine on the B-58.

For these tests we used the main runway at Edwards. My plan was to make a few idle power coasting runs to measure the airplane rolling and air resistance from its natural deceleration, a run or two to measure the drag chute forces (for which we put a strain gauge on the drag chute D-ring). I needed the drag chute data in order to subtract out its deceleration energy during our preliminary runs. We made several runs at increasing speeds to gather the necessary data for our final, big, test. We planned to investigate the use of aerodynamic braking to slow the airplane to the desired brake energy levels instead of using the drag chute. For the aerodynamic braking tests, Fitz would chop the throttles at the desired speed, then raise the nose to increase the drag—which that big delta wing had lots of—and then lower the nose and get on the brakes and deploy the drag chute, while I followed in the chase car. For each test, General Dynamics installed new brakes on which we did a low energy "burn-in" run, then parked the airplane at the end of the runway in the run up area and put on new tires for safety.

GD had installed "fuse plugs" in each wheel that would melt if the temperature from the brakes got too hot. This would release the air pressure from the tires and prevent them from exploding. If a person stood too close or in the wrong position, the force of the exploding tires could kill him. These fuses normally "popped" about 10 minutes after a moderate energy test run as the wheels heat-soaked, and we never had a tire explosion.

Even though GD had built steel "fenders" over the wheels to protect the airplane from flying debris and fire–both of which we expected –I wanted more safety. Since we were only to accelerate and stop, without actually becoming airborne, I asked GD to replace the fuel with water, except for the small auxiliary tank, which held about 4,500 pounds of fuel, from which the engines were actually fed. At first, GD was reluctant, but finally agreed as long as we put a corrosion inhibiter in the water, which we did. The airplane serial number, 58-1020 was thereafter nicknamed "58-$H_2$0."

Wheel Fenders to Protect the Wing from Flying Wheel Parts

The test airplane was not an instrumented airplane, so we had no onboard data recording equipment. It was an operational airplane that was schedule to go into major overhaul immediately after our tests. Being an engineer, I wanted to know what was happening, so I asked Instrumentation Branch to install a pressure transducer on the nose gear strut, a strain gauge on the drag chute D-ring, and a small recorder. At first, General Dynamics was reluctant about messing with the nose gear, but finally agreed when the nose gear strut pressure transducer installation included a hydraulic fuse—I didn't know they made such things, but they did.

This nose gear strut pressure data provided a remarkable amount of useful information. Think about it for a moment, and the way your automobile reacts when you accelerate and stop suddenly. As you accelerate, the front of your car rises up a little, and as you stop, the front settles down. We could measure these reactions with the pressure transducer on the nose gear strut. As the engines were started and run to power with the brakes set, they pushed the nose gear strut down. The strut was compressed more and more as the engines were brought up to military power, and then maximum afterburner power. The stress was relieved when the brakes were released, and, as the nose was lifted off for the aerodynamic braking tests, the only reaction was the residual pressure in the strut with no load on the nose gear. The stresses returned immediately when the nose gear was brought back down on the runway, braking was applied very hard, and the airplane came to a stop with the nose bobbing up and down. Just like your automobile. Because of this simple piece of instrumentation, I knew everything that was happening on that airplane.

One day I noticed that the nose gear reaction was getting lighter, even though we had not changed the water loading. I called the crew chief and asked him to check the balance tank (a small fuel tank in the tail, used for center of gravity control). Sure enough, he found that the aft fuel tank valve was leaking water into the balance tank, shifting the center of gravity aft.

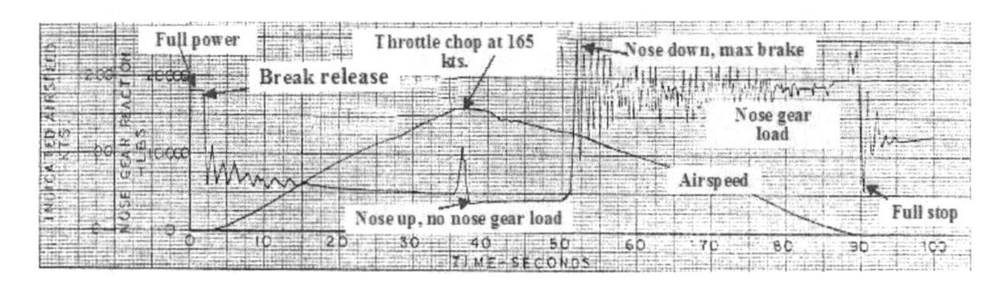

Oscillograph Traces Showing the Acceleration, Throttle-Chop, Aerodynamic Braking, and Final Braking to a Full Stop

On another occasion, the nose gear didn't load up immediately, as it should have, when Fitz put the nose down. I called Fitz.

"Why didn't you streamline the elevons after you put the nose down?"

"How did you know I forgot to do that! You don't have the elevons instrumented."

"Don't ever try to fool your flight test engineer!"

The end of November was getting cold. A couple of nights were freezing. Oops! I called the GD representative.

"Don't we have the aft fuel tank 95 % full of water?"

"Yes we do."

"When water freezes, doesn't it expand 10%?"

I had never heard a grown man scream over the telephone before. We both had the same nightmarish vision of waking up some cold morning to find this $25 million airplane sitting forlornly on the ramp, with its back split open like some giant, abandoned aluminum cicada husk. Thereafter, recording the water temperature was part of our pre-fight checks, and the water temperature was down to 38° F by the time of our last test.

On our first, moderate energy test run, a tire started smoking about ten minutes after we stopped, and then burst into flame. The fire department, always standing by for these tests, immediately extinguished the fire, but the fire should not have happened at that energy level. Therefore, I kept the airplane on the main runway until the wheels had cooled enough to examine them closely to see why the problem.

Colonel MacIntosh, the base maintenance officer, didn't like my keeping "his" main runway closed for four hours on a good flying morning. He accepted my decision, but threatened: "Pack your mukluks, Lieutenant Ransone—if this happens again, you are off to Thule AFB in Greenland!"

So it was with some trepidation when I had to report to his office and "chew him out" for standing in the wrong place after another test run.

I had studied the hazard zones around a tire that explodes. I found from the laboratories at Wright-Patterson AFB that when a tire explodes, it explodes radially in line with the tire, or laterally along the axle line. The safest area is within the 45-degree wedges between those boundaries. I advised Colonel MacIntosh that he had been in danger, and told him the safe areas.

"OK, thanks," he said, with a twinkle in his eyes. "You can unpack *one* of your mukluks."

The day of the big test arrived. It was the last chance we had, because the airplane was scheduled to go back into service. I had put together the test card that called for maximum power acceleration to 160 knots, which was the speed at which the plane could safely continue the takeoff on three engines. Fitz was to pull the firewall shutoff to stopcock all four engines and immediately raise the nose for the aerodynamic braking. He would then lower the nose at 140 knots and start maximum braking to a full stop with the anti-skid. All this without using the drag chute. This would demonstrate the 18 million foot pounds per brake of braking energy, and the ability of an Air Force pilot to stop the airplane from the three-engine takeoff speed without using the drag chute.

It was about 5:00 pm, after the day's flying was done since we were about to close the main runway for several hours. All of Edwards' brass was on hand to watch.

"Where should we stand?" Asked General Branch, the center commander.

"At 13,500 feet," I answered with confidence.

Then, in typical Hollywood fashion—***TROUBLE***! The ground power cart needed to start the jet engines wouldn't start! Fitz called for another one. The sun was going down. The photo-theodolites, needed to record the test data, had only a few more minutes of daylight.

The tower radioed: "The cart is coming! It just passed the maintenance hanger."

The sun kept sinking.

Getting darker.

"The cart just passed base ops. It's coming!"

Getting darker.

"The cart's at the plane!"

Almost too dark for the cameras to record the test data. Only a few more minutes of daylight left.

Engine start!

Quick preflight!

Taxi into position. Cameras ready. Fire trucks—every fire truck on the base—ready, red lights flashing.

The bosses waiting expectantly at 13,500 feet.

In the chase car–we're ready.

Fitz's countdown: "Military power… Check. Afterburner light off… Check. Maximum power...now… Check. 5..4..3..2..1..brake release."

With the last orange rays of the setting sun glinting off its silver sides, white hot flames blazing from the four powerful afterburners, the beautiful plane thundered down the runway, accelerating rapidly.

“160 knots…Engine chop.. .nose up.”

In the chase car, racing down the runway behind—watching for anything amiss. Everything looks good.

“140 knots...nose down. Full brakes!”

Fitz brought the plane to a smooth stop, with tires bursting into flame, fuze plugs popping, fire trucks roaring up and their crews scrambling out to foam the burning wheels, and Fitz clamoring down the escape rope—all within less than an airplane’s length of the general!

God! What a sight!

And the bosses coming out to greet us, not congratulating Fitz, who did the actual test, but demanding to know “where is the engineer who planned this?”

It could have been even more spectacular: As the plane came to a stop, the aft fuel tank high level shutoff valve failed, and the tank vented large quantities of liquid, cascading down over the wing, hissing onto the blazing wheels.

Quantities—not of fuel—water.

H_20.

"You Had Your Hand On The Stick!" "No I Didn't!" "Yes You Did!"

After a B-58 disintegrated over southern Oklahoma during a Mach 2 engine shutdown test at 36,000 feet, killing its two-man crew, several changes were made in the B-58's flight control system. The initial structural failure to that airplane, caused when it exceeded the 1 ¾ -degree sideslip limit for Mach 2 at 36,000 feet, was that the aft fuselage was literally twisted off from the side loads on the vertical tail. It was like someone had grabbed a child's airplane and twisted off its tail. Less than two degrees sideslip doesn't sound like much, but, believe me, at Mach 2, where the dynamic air pressure on the airplane was over 1,200 pound per square foot, the side forces had us literally hanging against our shoulder harnesses.

For this next story, you need to know a little more about the B-58's flight control system. I'll try to keep the explanation straight forward.

To counter an outboard engine failure at Mach 2, where the directional stability (Cn_b, for you engineers) is getting pretty low anyway, a very large vertical fin and rudder were needed. Because of its height above the airplane's center of gravity, large rudder deflections caused a considerable rolling moment, which was automatically counteracted by opposite aileron through a mechanical linkage called the Aileron-Rudder Interconnect (ARI).

The high vertical tail causes rolling moments when the rudder is deflected to counter yaw

But these weren't normal ailerons, they were "elevons"

mounted along the trailing edge of the big delta wing. They provided elevator functions collectively with up and down control, and aileron functions differentially for roll control. Since this was before microcomputers could make barn doors stable, some up elevon was needed for longitudinal stability. Aileron movements resulted in more deflection on one side and less on the other side, which caused differential drag, which resulted in yaw due to aileron, which required rudder deflections to counter, which…any way, you get my drift. All of this was handled automatically through the ARI. In addition, the airplane had special rate dampers to automatically counter roll, pitch, and yaw movements and help the pilot fly the airplane safely. Without these dampers, the airplane had a serious lateral-directional instability problem. Similar to your automobile when it has really *really* bad shock absorbers: The car rolls from side to side, and it is hard to keep it pointed in a straight line. This is also called "Dutch roll," and got this name from the motions of a Dutch ice skater, leaning forward, with his hands folded behind his back, weaving from side to side as he skates across a frozen pond. This is an excellent and graphic analogy.

This lateral-directional instability was caused by the configuration of the delta wing itself. Delta wings have an inherent dihedral effect, which can exhibit lateral-directional instability if the dihedral is very large, which characterizes a delta wing. You may note in the tail-view picture that the B-58 had no physical dihedral—the wings are straight out from the fuselage in order to minimize the dihedral effect.

The ARI ratios were changed as a result of the Oklahoma accident, so these changes had to be reevaluated. Also, the airplane was placarded to prohibit supersonic flight with dampers inoperative. Fine. Don't go supersonic without your dampers, but what if they fail at Mach 2? How do you get down OK?

Re-enter the steely, blue-eyed team from Edwards—Fitz and me. Once again we were flying out of GD in Fort Worth. (The fact that my mother lived in Fort Worth had nothing to do with all of this TDY!) Our mission, in September of 1960, was to evaluate the effects of the new ARI ratios on handling qualities, and to demonstrate that the airplane could be decelerated safely from Mach 2 with all dampers off, for which we had to develop the technique.

With dampers on, all pulses were essentially deadbeat. Good. That's what dampers are supposed to do. Dampers off, however, was more sporting. With the yaw damper off and the roll and pitch dampers on, at Mach 1.2, lateral-directional damping took 9 cycles and 24 seconds to half-amplitude. The roll damper inputs were driving the oscillations through the ARI.

At Mach 2, with roll and yaw dampers off, the oscillations started going divergent in two or three cycles. Bad news. But on analyzing the oscillograph records—back in those days, we had to read the oscillograph records by candle light (just kidding!)—I noticed that the lateral stick movements, although very small, were leading the airplane oscillations. I asked Fitz why he didn't let go of the stick.

"I did let go."

"No, you didn't!"

"Yes I did…but I kept my hand around it so that I could grab it quickly,"

"Your body motions, driven by the airplane movement, are bumping your hand against the stick enough to drive the oscillations."

The next time, he held his hand farther away from the stick, and, sure enough, the airplane motions damped out. But slowly, very slowly.

At supersonic airspeeds, with dampers off, the oscillations were very easily excited and very lightly damped, and manual attempts to control them could easily lead to in-phase coupling and loss of the airplane and crew. In trying to fly the airplane with the roll and yaw dampers off, several times Fitz had to turn the dampers on quickly in order to recover the aircraft from its divergent lateral-directional oscillations. Very quickly! On one test we were about one second from loss of control (see the picture on the next page).

The lateral-directional motions exhibited large roll and small yaw excursions, which suggested using the ailerons to damp the oscillations. Bad choice. The yaw due to aileron drove the divergence. Fitz found that he could damp the oscillations if he ignored the large roll oscillations and concentrated on controlling the small yaw motion with the rudder pedals. It worked. No yaw, then no Dutch roll. But Fitz found a better method—just put in a small amount of sideslip, which had a strong damping effect. That is the method that was added to the flight manual.

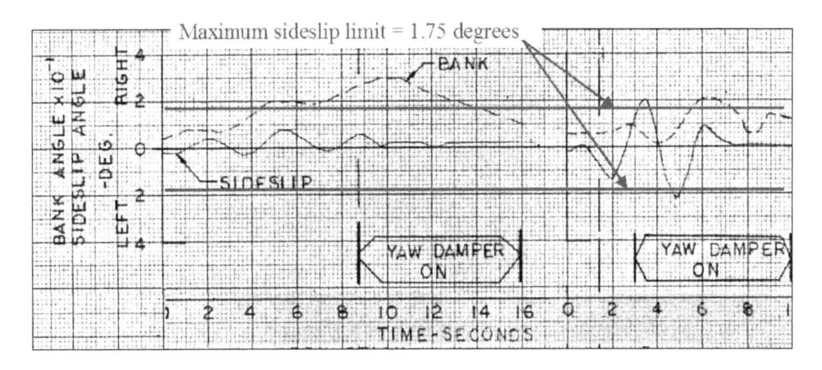

Dangerous B-58 dampers off lateral-directional oscillations

Note that if Fitz had not turned on the yaw damper during the second test shown, the airplane would probably have disintegrated in another second or two.

Fitz also found that, with practice and concentration, he could control the Dutch roll with aileron, but that method was too tricky for mere mortal pilots. The oscillograph records substantiated what Fitz had demonstrated.

Our big test came at Mach 2, when Fitz turned all dampers off flew straight and level for a few seconds, and then reduced power to idle for a deceleration to subsonic speed. I'm here to tell you that it worked fine. The only aircraft perturbation was when we decelerated from supersonic to subsonic flight, and the wing aerodynamic center shifted from the 45% chord forward to the 25% chord, the airplane nosed up slightly, which Fitz corrected.

In retrospect, I realize that this was a very hazardous test, probably one of the most hazardous I flew during my entire Edwards career. But I was completely absorbed in analyzing the data and solving the problem. I was confident of my data analyses, had complete trust in Fitz's ability, and never really thought about our safety.

We completed our mission successfully, returned to Edwards, and wrote our test report. We had solved the problem of how to return the airplane safely after dampers failure when flying at twice the speed of sound.

"There's a brown bear in the back seat!"

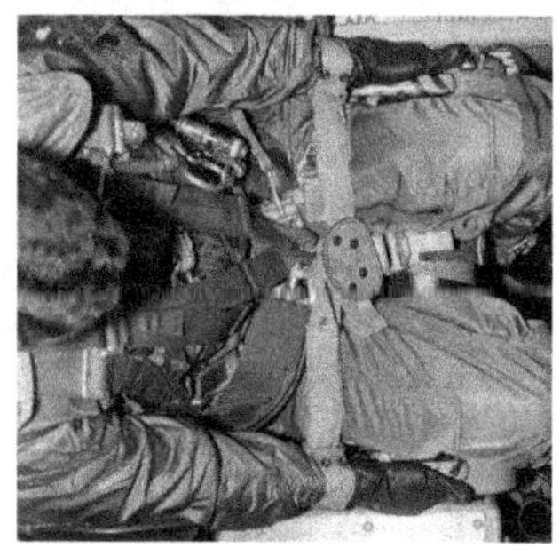

Strapping into my rocket ejection seat before a B-58 flight

My next B-58 test program was for the rocket escape capsule tests. There was a very serious problem with bailing out of the B-58 at supersonic airspeeds: you couldn't do it and live! During supersonic ejection it was a toss up (graphic imagery intended!) whether you would be sliced in half by the tall vertical fin behind you before the 1,400 mph slipstream wrapped your arms in knots behind your back and the 600-degree Fahrenheit ram air temperature fried your bones.

We had rocket powered ejection seats with several clever devices to help hold our body parts together until we slowed down. But with an air pressure of 1,200 pounds per square foot from the wind blast, and the searing 600° F air compression temperature, there would be no doubt of the outcome. When the B-58 disintegrated over Oklahoma, parts of the crew were scattered over a wide area, and it was reported that their orange flight suits were bleached pure white by the hot wind blast.

Strapping in for a B-58 flight was fairly complicated: We had long straps that we attached to our ankles, which were in turn attached to the floor of the cockpit. Upon ejection, as we rode up the seat track, these straps would pull our legs into the center and lock in order to keep them from wrapping around our necks in the wind blast, before the straps broke off as we left the airplane. We also had an upper leg restraint that crossed our thighs to further secure our legs. We had a seat belt and shoulder harness, with loops at the center buckle to put our wrists through. To eject, we would have to pull down our face plates on our helmets, put our hands through the loops, pull them out towards the seat arm rests, grab the handles tightly, pull up the handles to jettison the canopy, and then pull the exposed trigger to fire the seat. The straps would prevent our arms from flailing about and wrapping around our necks. But since 22% of the B-58s had crashed, and crew survival rate was only 44%, something obviously had to be done.

The new idea was to place each of the three crew members in a small capsule that would close around them upon ejection, thus protecting them from the wind blast. The parachute was attached to the capsule, and the crew stayed in the capsule all the way down and landing.

The capsule had been static tested but still had to be sled tested and "man rated." The sled tests would be conducted at Hurricane Mesa, in Utah. Upon assignment to this program, I flew to Hurricane Mesa to witness the first capsule test. Two pilots flew me to Utah in a small, twin engine Beech airplane, we landed on top of a mesa where a lieutenant met us and drove us to the test area.

The test track was on top of Hurricane Mesa, a flat "mountain" with a precipitous 1,400-foot drop at the edge. The test track was a "railroad" 12,000 feet long, with the rails mounted securely on a solid concrete ballast (base). The sled was composed of two parts: the motor part and a test vehicle attached in front. The motor part had a large, rocket motor in back that was used to accelerate the sled rapidly to high speed in only several hundred feet. The speeds—in the order of 500 mph or so—were too fast for normal railroad wheels, so the sled was supported on "brushes" with clamps holding the carriage to the rails.

The track ended only about 100 feet or so from the edge of the mesa. To stop the sled there was a water trough between the tracks near the end through which a scoop under the sled would drag in the water and stop the sled. The trough was built so that as the sled neared the end the water got progressively deeper, and the deceleration forces increased.

The operational concept was to set up the test with its instrumentation and high speed cameras, fire the rocket, and eject the test subject over the edge of the mesa. The test subject would fly over the edge of the mesa and would be tracked during its descent to the valley floor 1,400 feet below. The sled would hit the water "brake" and stop from 500 mph in about a hundred feet.

For the test that I witnessed, the objective was to test the rocket ejection capsule and also to see what would happen if a B-58 front windshield hit a 7-pound duck at 500 mph. The (dead) duck was stretched across the track at windshield height where it would hit the windshield just after the capsule had ejected.

I watched expectantly as everything was made ready. Then the countdown, and the rocket motor was fired.

Then everything happened so fast that I could hardly see it: With a loud roar and a cloud of smoke, the rocket motor accelerated the sled to 500 mph, the capsule rocket motor fired, the capsule blasted out of the cockpit, flew out over the edge of the mesa and the parachute opened, the windshield hit the duck, and the sled stopped in a cloud of water spray. All this in less than ten seconds! The sled was traveling 500 mph and it was less than 100 feet from the edge of the mesa! I watched, fascinated, expecting to see the sled fly over the edge.

But it didn't; It stopped safely.

I asked the lieutenant how many times you had to watch this before you could take your eyes off the sled and watch the test subject. He said that he had seen a half dozen or so tests, and that this was the first time that he had seen the test subject.

The capsule worked fine, but the impact with the duck was something else! The B-58 windshield was one inch thick tempered, laminated, safety glass, designed to withstand the high temperatures and wind blast of twice the speed of sound. Furthermore, it was slanted back at about a 60-degree angle, both from top to bottom and from side to side. It's one serious, tough piece of glass. Also, a sheet of stainless steel had been installed behind the capsule to protect the cockpit aft bulkhead from the rocket motor blast.

The duck took out the entire left windshield panel, punched a 9-inch hole in the stainless steel doubler at the back of the cockpit, went through the aft bulkhead of the cockpit, the bulkheads of the (empty) electronics compartment, and through the forward bulkhead of the second crew station!

Not good at all! Hitting a duck at 500 mph would obviously kill the entire crew.

Rocket escape capsule firing test at zero-speed and zero-altitude

The remark about the brown bear at the first of this story has to do with the man-rating program. It was determined that a brown bear has a back and spine that's very similar to a man's, much more so than a chimpanzee's. For this reason, it was decided to utilize a brown bear for the first live, "manned" capsule ejection test. This was, in fact, done successfully—and safely—with a sedated bear and under the watchful eyes of the SPCA and Air Force Human Factors specialists.

This was my last work on the B-58, and I did not witness these actual flight tests. I was transferred from the Bomber Branch to the Rotary Wing Branch, thus leaving my beloved B-58s and entering a long association with Vertical and Short Takeoff and Landing aircraft.

But before I leave Hurricane Mesa, I have to tell you a story of a sled test film that I saw in a Hurricane Mesa test report: This test was of a rocket ejection seat test with a chimpanzee. The film shows the keepers in their white uniforms strapping the happy chimp into his ejection seat. The chimp is chattering excitedly, grinning from ear to ear, looking around, and thoroughly enjoying all of the attention.

The keepers hand him a banana.

Chimp happily peals the banana and sticks it into his mouth.

Ka-BLAM! Rocket fires!

The sled roars down the track accelerating to about 500 knots in only a few seconds!

Ka-BLAM! Rocket seat fires!

Chimp is ejected high into the air and sails over the edge of the mesa, spread eagle, arms, legs, and tail flailing, tumbling end over end!

Parachute opens!

Chimp floats gently down to the valley floor 1,400 feet below.

Chimp lands safely.

Two white uniformed keepers rush up to him and hand him a banana.

Chimp takes one look at that damn banana, and then throws it as far as he can!

"Isn't He Flying Awfully Slow For A C-130?"

In the fall of 1962, Air Force Systems Command (ASD) at Wright-Patterson AFB and the AFFTC conducted a series of tests to determine the minimum field surface conditions on which assault transport operations could be performed. A follow-on to an earlier "Rough Road" test of the Lockheed C-130B and Fairchild JC-123B transport airplanes, this new project was called "Rough Road Alpha," and included two C-123 and three C-130 aircraft.

The C-123 fleet comprised a standard aircraft, and a modified aircraft on which a CJ-610-1 turbojet engine was installed under each wing in addition to the two R2800-99W supercharged radial engines and propellers. The C-130 fleet comprised a standard C-130, a slightly modified C-130B with higher propeller speeds and larger flap deflections, and an NC-130B that was highly modified by addition of boundary layer control (BLC) blowing. Since I seemed to get the weird tests at Edwards, I got the BLC airplane.

NC-130B was a production C-130B, powered by four T56-A-7 turboprop engines, and on which strange looking jet engines had been hung under each outboard wing section. They were strange looking because, although they had air inlets in the front for air to get in, they had vanes on the back that could close off the jet exhaust so no jet exhaust could get out!

NC-130 BLC airplane. Note the jet engines under the wingtips and the large flap deflection and rudder deflection. Jet exhaust valves are open

One might think the aircraft would puff up, sort of like a blimp, and pop like a balloon.

Not to worry. The jet exhaust was ducted up through the airplane, and eventually came out over the flaps (which could be deflected to 90 degrees—the jet airflow followed this large deflection because of the coanda effect), the ailerons and elevators, and the rudder (with increased deflection and lengthened chord). The photo shows the jet engine exhaust open for dumping the BLC lift after landing in order to load up the landing gear for stopping.

I first heard about this program one Friday night, just before dinner, when Ray Johnson, the Chief Project Engineer for Rough Road Alpha, telephoned me. He described the program, and asked if I would run the BLC tests.

"OK. When is the airplane coming to Edwards?"

"It just landed," Ray answered, smugly. "Your first flight is in the morning at 0900."

"Thanks for the advance warning. Got anything for me to read tonight?"

NC-130 BLC crew and airplane. I'm the second from the left, kneeling on the front row.

Bright and early the next morning, I arrived at the airplane, met the pilot and copilot from ASD, and was accosted by the crew chief.

"How much tire pressure do you want?"

"55 psi." I answered, also smugly—thank you, Ray, for the Rough Road test reports last night!

Our first tests, airspeed calibrations, showed that the pitot-static system's improvements had not kept up with the airplane's aerodynamic improvements. C-130B assault landing approaches were normally flown at about 93 knots and touchdowns at about 87 knots. For the NC-130B BLC airplane, these speeds were reduced to about 70 and 65 knots, respectively.

The problem, however, was that we really had no idea how fast we were flying. With 90-degrees of flaps and BLC air blowing over them the airplane did not stall but just developed higher rates of sink. As we "decelerated," the indicated airspeed slowed to 55 knots, then 50 knots, and then started increasing to 55 knots, and then 60 knots, and the sink rate increased and the pitot static errors increased. We believed that we were slowing down, but the airspeed indicator disagreed.

On one 70-knot approach to the main runway, an F-104, cleared for landing behind us, finally called the tower:

"Tower! Isn't that airplane going *awfully slow* for a C-130?"

"Ah, Eddie Tower. Would you send a big tow truck out to Harper's Dry Lake? Our airplane is stuck in the mud."

The first assault field operations we tested were on soft clay at Harper's Dry Lake. U.S. Army Corps of Engineers from the Waterways Experiment Station, Vicksburg, Mississippi, were supporting our tests by measuring the load bearing capability of the test fields. They used California Bearing Ratio (CBR), in which a rating of 100 is equivalent to crushed limestone. These engineers would press a probe down into the surface and measure its resistance. They did this for areas of Harper's Lake, and we would land and takeoff at those areas, and note any problems.

The problem was, there weren't any problems! We should have been sinking into the soft clay, but we weren't. We theorized that the dry, sun-baked clay on the surface was being pressed into the damp clay underneath by the wide tires, and we were actually creating a workable runway surface as we went.

On a Sunday morning, we were at Harper's Lake, taxiing our big NC-130B slowly around the lake bed, following the engineer as he walked along, sticking his probe into the soil to measure its CBR.

An unusual sight, even for Edwards.

All of a sudden, we got the data point we were after! The airplane just sort of leaned over to the left and stopped.

Our left main landing gear had sunk into the soft clay, rutting 22 inches deep. The left wing tip was at eye level.

We radioed Edwards. Were they glad to hear from us on a Sunday afternoon?

Sure!

NC-130 BLC stuck in the mud at Harpers Lake

Why Land On A Hard Old Runway, When The Sand Is So Nice and Soft?

After two big bulldozers towed our airplane out of the mud at Harper's Lake, our next test area was in soft sand down at Yuma Naval Air Station, Arizona. The CBR of the sand test area, alongside the main runway, was zero. Zip. Nada. Nothing. The probe just went down into the sand with no resistance or reading at all. But taxi tests showed that the fat tires provided sufficient floatation for operations, even for this condition.

My main concern, however, was sand recirculation into the engines. With 90-degrees of flaps, and boundary layer air blowing over the wings and flaps, the sand would re-circulate under the wings and back up into the engine inlets during reverse thrust on landing roll out. This would not be good on the engine compressors, and the engines had no sand separators. Furthermore, our assault takeoffs were about 35 knots below the engine out control speed, and I didn't fancy flipping upside down right after takeoff. Neither did the rest of the crew. Since there was no sand ingestion problem during takeoff, I planned our tests to complete all of the assault takeoffs from the sand first, landing on the concrete runway, and then doing our assault landings in the sand, taking off at safe, higher speeds from the concrete runway.

I got no arguments from the rest of the flight crew, and we finished our takeoffs without incident.

The landings, however, were spectacular.

NC-130 BLC on landing rollout in soft sand at Yuma Naval Air Station

During landing rollout, with full flaps and reverse thrust, the entire aircraft was enveloped in a dense sand storm, with only the wing tips sticking out of the cloud. The pilot had to complete the landing rollout on instruments.

On one landing, a Navy A4D had been cleared to land on the runway. Shortly after his clearance, the tower cleared us to land in the opposite direction.

Things like this do not go unnoticed by Navy pilots.

"Yuma tower! Did you just clear that Air Force one-thirty to land on the other end of my runway?"

"Affirmative. But he isn't going to land on the runway—he's going to land in the sand beside the runway."

A moment of silence.

"Say again, tower. He's not going to land on the *runway*, he's going to land in the *sand* beside the runway?"

"That's affirmative."

Navy took a wave off—he didn't want any part of that crazy Air Force operation.

That evening we enjoyed drinks and dinner at the Yuma NAS Officers' Club. This club was a bit different from our Edwards Officer's Club. First, there was a large glass jar of whiskey above the bar with a dead rattlesnake in it. If you were game you could buy a swig. We weren't, and we didn't.

Another difference was a large, brass bell at the end of the bar. Navy protocol decreed that if you entered the bar with your hat on, you had to buy a round for the entire bar.

That night, suddenly we heard the bartender clang the big bell, and looked to the door to see a petrified young ensign standing there with his hat on. He took one look around the crowded bar, instantly realized that he was about to lose his entire month's salary, turned, and bolted from the room.

He didn't furnish us with a round of free drinks, but at least he furnished the whole bar with a round of laughs!

We finished our tests without further incident, loaded the ground crew and photo theodolite cameras on the airplane, and prepared to fly back to Edwards.

Takeoff position on the runway. Full power check. Ten seconds, OK. Then two "bangs" and number three engine quit cold.

I have never heard a turboprop engine backfire, before or since.

We tried again.

Same result.

The engine would not hold takeoff power, because the sand had eroded the compressor blades enough to cause a compressor blade stall and instantaneous power failure.

We flew back on another C-130, and left the NC-130B BLC at Yuma for an engine change.

Good plan to do our assault takeoffs first.

Vertical Flight — The VTOL Test Stand

Perhaps the most unusual aircraft to come to Edwards in the foreseeable future would be the VSTOL, Vertical and Short Takeoff and Landing aircraft. They were referred to as "aircraft" rather than "airplanes" because the "-plane" part of the latter definition referred to the aircraft's wings that generate the lift needed for flight. These new VSTOL aircraft didn't always rely upon wings in order to fly, so were categorized in the broader definition of "aircraft." They included helicopters as well as a wide variety of more innovative propulsive lift concepts.

During the 1950s, the Air Force, Army, Navy, and NASA were very interested in this VSTOL capability. They wanted the high speed and efficient flight of fixed wing airplanes with the short, or "no" runway needed and hover capability of helicopters. There were many concepts and designs proposed, many built, and some even flight tested. One of the first to be flight tested at Edwards, that was not an actual helicopter, was the Hiller X-18.

The X-18 started life as a World War II glider, that was later modified into a high wing, twin-engine, propeller cargo transport aircraft called the Chase YC-122C. Hiller added a special wing that could be tilted up to 90-degrees (vertical). Two very powerful turbo shaft engines (from earlier VTOL experimental aircraft) were added, one under each wing. These engines had two power sections mounted side-by-side, geared together and driving contra-rotating propellers through a common gearbox. In order to keep the wing from stalling during slow speed flight, the wings were shortened so that they did not extend past the propellers, which kept them immersed in the high velocity propeller slipstream.

Flight controls for normal (airplane) flight were conventional—aileron (roll), elevator (pitch), and rudder (yaw).

For vertical flight, with the wing in the vertical position, height was controlled by the throttles, roll was controlled by differential power between the right and left engines that were pointing straight up. Directional (yaw) control was through the ailerons at the wing trailing edges, in the strong propeller slipstream. Pitch control was a different matter, altogether: A small, jet engine was mounted in the aft fuselage with its exhaust exiting the rear of the aircraft. A diverter vane was installed at the end of the exhaust pipe, which could be diverted either up (nose up) or down (nose down) for pitch control.

With its boxy glider-type fuselage, short clipped wings, and big jet exhaust pipe hanging out of its tail, it was probably the ugliest aircraft ever to come to Edwards!

Ground tests were conducted at Hiller's plant in Palo Alto, California, and then the aircraft was brought to Edwards for flight tests over the dry lake, where emergency landings could be made rapidly in case of emergencies, which were clearly anticipated.

Hiller test pilots George Bright and Bruce Jones flew the very short flight test program which demonstrated more than anything else, how *not* to design a VSTOL aircraft!

Two main problems: (1) throttle control was not fast enough for either height or roll control during vertical flight, and (2) cross-shafting, between all engines, is imperative for flight safety in order to keep all propulsion systems thrusting in the event of an engine failure.

After only 20 flights and several hair raising incidents, the Air Force terminated the flight test program on the grounds that: "...further flight testing of this aircraft will undoubtedly result in a catastrophe!"

Charlie Crawford, head of Rotary Wing Section of Flight Test Engineering at Edwards, had an idea: Why not build a special test stand on which the aircraft could be safely mounted, the engines run, and the resulting forces on the aircraft and the propeller wash ground effects measured. The aircraft could be raised and lowered by hydraulic rams up to 15 feet (essentially out of ground effect), and the aircraft could be oriented in various angles of pitch and bank. This research would also support the forthcoming Vought-Hiller-Ryan XC-142A Tilt Wing VSTOL Transport being developed and built at Vought's Dallas plant.

Charley presented this idea to Wright-Patterson AFB, which agreed, and provided the necessary funding.

Now, who in Flight Test Engineering should be assigned to this weird program?

Rob Ransone, of course!

It sounded really interesting, so I readily accepted, and was given the assignment.

The technical specifications for the VTOL Test Stand had already been prepared, the Request for Proposals had been released through the Edwards contracting office, a contractor selected, and the contract signed for design, fabrication, installation, and proof testing. The contract also included a special steel cradle on which to mount the X-18 on the VTOL Test Stand.

I researched the project and reviewed the specifications. I discovered that the design of the test stand tilt mechanism was such that very complex calculations would be needed in order to transfer the data reference coordinates to the aircraft axes from the test stand axes. The first thing I did, then, was to tell the contractor to "turn the tilt head upside down so that the data measurement axes are the same as the airplane axes on which the forces are being measured." We could then read out the forces directly on the oscillograph recorders as we monitored the tests. This was before microchips. Otherwise we would not know the aircraft forces until we ran the data through a mainframe computer program.

The VTOL Test Stand was finally built, installed, and static proof-tested.

My approach to flight tests was always to identify what could go wrong, the likelihood that this could happen, and implement suitable precautions. This involved anticipating the things that could go wrong, what indications I would have that something was amiss, and what I could do in advance to prepare me for overcoming the problem with the resources I might have on hand at the time. In this case, I was concerned about the dynamic forces on the stand with the aircraft engines and propellers turning.

Although I knew only a little about structural dynamics, I did know that the spinning propellers on the X-18 would vibrate the test stand. I also knew that the test stand would have varying natural vibrating frequencies that would be high when the stand was near the ground and the exposed hydraulic rams were short. The natural frequency would become lower as the stand was raised, much the same as high-pitched organ pipes are shorter than the low frequency ones. I knew that as long as the structural rigidity of the hydraulic rams damped the vibrations there would be no problem.

But I also knew that when solders march down a road they break step when crossing a small bridge because if their synchronized footsteps happen to be in phase with the natural vibrating frequency of the bridge, the bridge can collapse.

I called the Flight Dynamics Laboratory at Wright-Patterson AFB and asked them to come to Edwards for the first X-18 tests.

They refused, on the grounds that "…if we are there, and something bad happens, then we could be held responsible!"

I was surprised and disappointed with this response. What did they expect me to do? Cancel the program and admit that we had wasted all that money on a test facility that we were afraid to use? Or just press on and maybe have the test stand go into resonance with the X-18 propellers, the airplane break off, drop 15 feet, and kill the pilot? Neither of these options was acceptable, so, since I was assigned responsibility for this project, it was up to me to figure out how to make it work safely. This is what the Air Force was paying me to do. I don't remember talking to anyone else at Edwards because I didn't know anyone else who knew any more about this than I did. I thought that if anyone could figure it out, I could. This was my job and I was confident that I could figure it out myself. I had no trouble "thinking outside the box," and, in fact, I suspected that this had been noticed by my superiors, and was why I got these weird assignments. This was fine with me, because I enjoyed new challenges that required "thinking outside the box."

I had met Carl Horst, from the Flight Dynamics Lab, and we had discussed the likely vibrating characteristics of the test stand. He knew a lot about vibrations and natural frequencies, and amazed me with his stories of wandering through the woods in the winter and pushing over dead trees—some up to eight inches or so in diameter. He would find a dead tree, and push against it at shoulder level and then release his pressure, noting how the tree leaned ever so slightly. After the tree had sprung back, he would push again, keeping in phase with the oscillations of the tree. As the oscillations grew, he pushed harder and harder, and after a few minutes the tree would be swaying back and forth so violently that it would topple over!

(Years later, I tried this myself, and pushed over a dead tree that was about six inches in diameter. What fun!)

So as these thoughts went through my mind, I realized that I could measure the natural vibrating frequencies and damping ratios of the test stand by pulsing it at various heights, and by comparing these frequencies with the forcing frequencies of the X-18 propeller. I could then determine the critical heights where the test stand would be in phase with the X-18 and proceed with great care at those heights. The damping ratios–cycles to half amplitude–would tell me how resistant the test stand was to the plane's vibrations and how safe the tests could be conducted.

On the head of the test stand I had Instrumentation Branch mount small instruments, called "accelerometers," that would measure the vibrations of the test stand. Then I set about to pulse the stand to measure its natural frequencies and damping ratios.

To do this, I got a spool of nylon parachute cord, and tested its breaking point in our laboratory. It broke at 125 pounds, after stretching 50%.

Next, I built up loops of this cord with enough strands to break at 1000, 2000, and 5000 pounds total load. I tied one end of a loop to a tug (used to tow airplanes around the base) and the other end to the top of the test stand. I also tied a long line to the break line so that when the stretched line broke it would not fly back and hit the tug driver.

I planned to set the test stand at various heights, pull the loops with the tug until they broke, which would pulse the rams, and measure the test stand natural frequencies and damping ratios on the oscillograph. By knowing the forcing frequencies of the X-18 propellers, I would therefore know the critical heights at which the test stand would be in resonance, and where to be especially careful.

I started out conservatively: First we tried it with the 1000-pound test loop. Nothing. Even the 2000-pound loop failed to pulse the rams enough to measure. The 5000-pound loop was a different matter. It was not a huge pulse, but it was enough to excite the rams' vibrations for accurate measurement. So we continued pulsing the rams at each one-foot height with 5000-pound loops until we had an accurate picture of the rams' natural vibrating frequencies from just off the ground to full extension. I had determined that the rams had high damping ratios and that their resonant frequency with the X-18 propellers was at 9 to 10 feet. Now I was ready to mount the X-18.

The big day arrived. We had the X-18 mounted securely in its cradle on the VTOL Test Stand. George Bright was in the cockpit, the base fire trucks on site, and everyone briefed. It would be a simple test, basically just engine runs at various heights to check out everything.

George started the engines, and we raised the stand a foot at a time, recording the loads and vibration data. I knew that the resonance frequency would be reached at nine feet height, and, sure enough, the vibrations increased slightly at that height. So we raised the stand only 6 inches for the next test.

X-18 on the VTOL Test Stand

"What's happening? Why just 6 inches? Is everything OK?" George wanted to know!

"Fine. Just a precaution," I answered over the radio.

The next 6-inch increment showed no increase in vibration, and the next increment resulted in lower vibrations, so the critical height was passed, and we completed the test safely. I had overcome the Flight Dynamics Lab's refusal to support me, and had figured out a way to complete these initial tests safely. I was very pleased with my solution to the lack of support from the experts. It proved that I could complete my job safely even without outside support. I was not dependent upon support from people beyond my control. I could do my job.

We completed this run, but our program was out of money, so this was the end of the program. I had the test instrumentation cut out of the aircraft with side cutters in order to preclude the aircraft from every flying again, and eventually it was scrapped.

But the X-18 had served its purpose: It had shown how future tilt-wing VSTOL aircraft should be designed, and the VTOL Test Stand was used successfully on a number of VTOL aircraft and helicopters after that.

The main thing that this program did, was get me involved in VSTOL aircraft, which I pursued for many years after that, both at Edwards and at American Airlines.

What Do You Do In A Helicopter When The Rotor Breaks?

While the VTOL Test Stand was being built, I had some free time, and my boss, Charlie Crawford, Chief of the Rotary Wing Section, decided that it would be good for me to get some helicopter experience. It sounded like fun, and I was assigned to a retest of Cessna's pretty little YH-41 helicopter. Again, the prototype "Y" designation indicated that the helicopter was no longer experimental, but was not yet certificated for production and sale.

In the late 1950s Cessna decided to get into the helicopter business, and hired Charlie Seibel, who knew a lot about helicopters, to design a small four-place, single rotor helicopter for Cessna, who knew nothing about helicopters. Cessna's instructions to Charlie were to "design us a helicopter that looks like a Cessna airplane."

He did, and it was a pretty little four-place, single-engine, single-rotor helicopter. His design was powered by a Continental FSO-526 supercharged reciprocating engine, de-rated to 270 horsepower for increased reliability, which was mounted horizontally in front of the cabin, with the transmission and main rotor shaft between the pilot and copilot.

Bob Baldwin and me (right) with the Cessna YH-41 helicopter

The AFFTC had flight tested the YH-41 for possible U.S. Army use, and reported in 1960 that it had good performance but unacceptable handling qualities. It tended to yaw and flip up on its left side immediately after an engine failure or throttle chop. Not good, since that is when the pilot needs precise control in order to start autorotation and a safe power-off landing.

Cessna had improved the handling qualities and installed a mechanical, leaf-spring roll rate damper that solved the control problem after a throttle-chop. But the Army decided it did not want the YH-41 because the new technology turbo-powered helicopters had much better performance. Cessna wanted to market the aircraft commercially, so the Army and the AFFTC agreed to re-evaluate the chopper as a courtesy to Cessna. This re-evaluation included a short variable stability program during which different springs were tried out in the roll damper to vary the roll damping. Bob Baldwin was assigned as test pilot, and I was the flight test engineer. We flew most of our flights during the winter of 1962, and I was delighted to fly this because I had never before flown in a "chopper."

Delivery of the YH-41 was not without humor: George Bright, a helicopter test pilot who had flown the X-18, was hired to ferry the YH-41 from Cessna's Wichita, Kansas, plant to Edwards. His flight plan was simple: basically to follow U.S. Route 66 at about 5000 feet altitude to California. En route through New Mexico, he was getting a bit low on fuel, and decided to stop at a filling station for gas.

This was a small helicopter, remember, so he could do that.

George landed next to the station, and attached small wheels to the landing skids in order to roll the chopper up to the pumps. The wheels were installed by using a long lever to lift up the skid and slip the wheel's axle through a hole in the skid. Normally this was easy, because the chopper weighed only 3100 pounds.

Except this time, he didn't have the lever properly inserted, and the chopper started slipping. Without thinking, George reached under the skid to catch it. Bad move. Although George's hands were in a slight depression in the macadam road and he was not injured, both hands were stuck under the skid.

“Excuse me,” he called to the station attendant, who had been watching all this in amazement. “Would you mind taking this helicopter off my hands?”

The rest of the ferry flight was uneventful, but not so our flight test program.

One of the YH-4 l’s stability problems had been “rotor blade weaving.” This was where the rotor blades vary their pitch slightly, during their rotation, because of changing aerodynamic loads on the blades. This caused a bad vibration of the aircraft at a frequency of twice the rotor rotation speed. Charlie Seibel had installed a small, viscous damper—much like an automobile shock absorber—at the rotor hub, between the rotor blade roots. This worked well, except on one flight at South Base.

Bob and I were doing throttle chops and autorotational entries from about 5000 feet density altitude (helicopters are tested by density altitude, not pressure altitude). The procedure was to establish a stabilized hover, start the flight test data recorders, then Bob would give a five second countdown, chop the throttle to idle, count two seconds (required by the handling qualities specification to allow time for a surprised pilot to react to an unexpected engine failure), and enter autorotation. Autorotational entry was attained by shoving down the collective pitch lever and pushing the cyclic stick forward to drop the nose.

In case you’re interested, autorotation is not just wind blowing up through the rotor. The rotor is actually driven *forward* by the resultant aerodynamic forces on the blades. The combination of forward aircraft speed, descent speed, and rotor rotation speed provides a resultant blade lift/drag force vector that actually drives the rotor blades forward into the wind. The pilot controls rotor rotational speeds by carefully following a prescribed descent rate: too shallow and the rotor slows down and stalls, too steep and the rotor over speeds. The pilot uses the kinetic energy stored in the spinning rotor blades to arrest aircraft descent and flare for landing touchdown, which can be as gentle as a power-on landing.

The throttle chop went fine, and we established autorotation with no problems. Then Bob applied engine power to recover from the autorotation and climb back to our next test condition.

Suddenly, without warning, a severe vibration shook the aircraft.

Big, heavy, bad vibration!

$R_{EE}{}^{EA}A_{LL}{}^{Y}$ $B^{A}{}_{AA}A^{D}$ $VI_{BR}A^{T}I_{O}N$!!!

Bob radioed Edwards tower that we had an emergency and required a straight in approach back to the main base hanger area, about three miles away.

The main rotor shaft, between Bob and me, was really pounding. We both had on parachutes, and I remember looking out the big Plexiglas door and thinking: "It's got to get worse than this before I step out!" I was not yet ready to join the Caterpillar Club, so named because a bailout from an airplane is like a butterfly emerging from its cocoon and spreading its new wings.

There was a panel by my right foot that I opened to look at the drive shaft between the engine and the transmission between us. That shaft was bouncing up and down about three inches in the middle! More than even a helicopter should bounce! I was concerned that something structural would break!

Not a good thing! You really don't want the only thing holding you up to break off!

"Bob, I don't think you should try for main base—I think you should set her down right here. Right now!"

This sounded pretty good to Bob, so he radioed the tower that we were setting down at South Base, and re-entered autorotation. In autorotation, the pounding reduced considerably, but was terrible during the landing flare and touchdown.

Upon deplaning (de-helicoptering?) we noticed that the damper between the rotor blades had broken loose on one end. The combination of the blade weaving and the weight of the broken damper on one rotor blade caused the severe vibration, the drive shaft and transmission might or might not have held together much longer.

Analysis of the oscillograph records showed the two per rev vibrations to be from +3.6 to -1.8 g during flare and touchdown. This is major, major vibration that could have resulted in a structural failure of the entire rotor assembly and loss of the helicopter and us.

Damper installation, and broken fitting after damper failure

Strange as it may seem to lay people outside the flight test world, neither Bob nor I had any fear either during or after this incident. During the incident, we were too busy figuring out what to do and how to get the aircraft down safely to be frightened. In fact, the only concern I remember having for our own safety, after seeing the way the drive shaft was pounding, was just to get down as quickly as possible without overstressing the transmission system. We were still thinking clearly, calmly, and logically, because that is what it takes to respond to in flight emergencies like this. Panic would only exacerbate the situation. This was a "matter of fact" thought process of assessing what bad could happen and how best to prevent it. Some people "have the shakes" after a life threatening incident such as this. For example, after a near automobile accident many people will pull over to the side of the road to recover. Such was not the case for us, nor was it the reaction for most of the professional test pilots and test engineers whom I knew at Edwards. It was a special mind set that seemed to rationalize: "Well, that was close, but we got down OK–we learned from this and will be more alert next time. Not to worry."

And these incidents also made great stories at Happy Hour.

Most of the tests were fun. for lateral control tests, I would set up a small jig around my control stick. When it was set, I would tell Bob and point in the direction for him to move the stick. He would move his control stick to hit the stop I had set, and I would yank off the jig in order to get it out of the way for Bob to recover. We did this both laterally and longitudinally.

On one test, we decided to see what would happen if Bob just pulled the stick back a small amount and held it.

This one was really fun!

But first you need to know a bit about how single-rotor helicopters fly: with great reluctance, fixed wing pilots claim.

With only one main rotor spinning, there is a reactive torque on the helicopter that tends to spin the fuselage in the opposite direction. Designers install a variable pitch propeller at the tail, spinning in a vertical plane, to counter this torque.

Another phenomenon that needs explaining is "translational lift." When a helicopter is hovering, its thrust axis is straight up the rotor shaft, just like a propeller. But in forward flight, the swept rotor area acts like an airplane wing, and the center of thrust, now called "lift," shifts forward from the shaft centerline to the center of pressure of the swept rotor area, roughly 25% of the way back from the leading edge of the swept rotor area disk. This creates a nose-up pitching moment and provides very large lift forces that enable a helicopter to fly at much higher altitudes than it can hover. You can see this action when a helicopter takes off. The pilot will lift off vertically, then push the nose down and pull up on the collective lever to add power. The helicopter accelerates, pulled forward by the tilted rotor, and then suddenly the nose pitches up, which the pilot has to counter by pushing his control stick down. This is when the rotor enters translational lift.

Note also, that the pilot is not flying the fuselage on a single rotor helicopter, but is flying the rotor disk. His control stick, called the "cyclic control" tilts the rotor plane. Although the helicopter is hanging from the rotor, aircraft motions are about the helicopter's total center of gravity, not about the rotor. It's much like swinging a bucket of water from its handle—you cannot spill the water, even when you swing it over your head. Similarly, if you place a glass of water on the floor of a single rotor helicopter, the pilot cannot spill the water, regardless of how violently he moves the cyclic stick.

OK, enough of "Helicopters 101," now back to our special test:

With back stick, the aircraft slowed down, the tail propeller no longer balanced the main rotor's torque so the airplane yawed 90 degrees to the right, we had slowed below the translational lift speed so the lift shifted back to the rotor centerline and the helicopter's nose dropped, then we entered a dive.

WOW! Disneyland charges lots of money for rides like that, and we got it for free!

Bob decided that I should be able to land the helicopter if anything should happen to him during flight, so he instructed me on how to fly and land it. I brought it down to a hover just above the ground, from which I could easily have landed it, and then Bob took over for the rest of the flight.

What fun!

Cessna built only six YH-41s. By the time it had fixed all of the handling quality problems, more powerful and more reliable turbine powered helicopters were available, and reciprocating engine helicopters became of interest only to small operators with small budgets, or private pilots.

I saw Charlie Seibel years later at Bell Helicopter. He remarked: "I'm glad they quit flying those helicopters before someone got killed."

The reason that we were wearing parachutes is an interesting story in itself: Before a special test in the Kaman HH-43 helicopter, helicopter flight test crews normally did not wear parachutes because in case of an engine failure–although very rare, about the only in-flight emergency expected–the chopper could be auto-rotated to a safe landing just about anywhere in the desert.

The HH-43 was an unusual helicopter. It had a boxy fuselage, twin tail booms, and a complicated horizontal tail with double vertical fins on each end of the horizontal tail. It did not have a tail rotor to counter engine torque because it had two rotor shafts, angled slightly out, with two single blade rotors that intermeshed. It had terrific hover performance because of its very low rotor disk loading—pounds of helicopter supported by each square foot of rotor swept area. The USAF Strategic Air Command, as well as other Air Force units, loved this helicopter because, in the event of an aircraft crash, it could hover over the fire, blowing away the flames to clear a path for the crew to escape safely. These fires were too hot for other helicopters to hover over because the heat raised the density altitude.

Kaman HH-43* Husky *helicopter shown with red fire suppression equipment in the sling

But, like the YH-41 helicopter, it had a peculiar and dangerous flight control problem in the event of an engine failure and entry into safe autorotation. In order for the pilot to push the nose down the correct angle to develop the forward speed and descent needed to initiate autorotation, the big tail assembly had to change angle automatically and immediately. The special test on the occasion in question was to demonstrate a safe auto-rotational entry after a throttle chop to simulate a sudden engine failure. Since this was a hazardous test, both pilots, Jimmy Honacker and Gene Colvin, had on parachutes.

The test entry went fine: Jimmy set up the test condition at 5000 feet (which, by the way, was only about 2,700 feet above ground level here in the high desert), gave the countdown for throttle chop, counted the required two seconds that was required by regulation to enable the pilot to react to a surprise engine failure, then shoved the nose down to initiate autorotation.

That's when the horizontal tail failed to move, and the helicopter entered an uncontrollable dive. As it picked up speed the rotors over-sped and flew off! Jimmy gave Gene the command to bail out, jettisoned his door, stepped out, and immediately pulled the rip chord on his chute. Gene's door, unfortunately, failed to jettison, and he couldn't get it open. Just then because of the high dive airspeeds, the big windshield shattered. Gene put one foot on the instrument panel, leaped out through the opening, and yanked his ripcord as he went. His chute opened, he swung once, and landed face down in the desert sand. Fortunately both pilots were unhurt, but the HH-43 was, of course, destroyed. After that incident, all helicopter test crews wore parachutes.

Now for the humorous part of this adventure: Jimmy carried a small wire recorder (remember this was in 1963) in his flight suit, connected to his mike. This was a so called “hot mike” that would start recording automatically when Jimmy said anything, and recorded continually anything that Jimmy spoke into his microphone. Jimmy used these recordings to prepare his flight notes after each test flight, rewound the wire, and subsequent flight comments were recorded over the previous flights. Sometimes there were bits of earlier flight notes after the end of the current flight notes. This created the amazing recording that I am about to relate:

(Current flight:) “OK, Gene. Ready? 5…..4….3….2….1….chop…1….2…………….***GET OUT GENO! GET OUT!***……”

(From the previous flight – so help me:) “Well, now, that wasn’t so bad, was it?”

If I hadn’t heard the recording myself, I would never have believed it!

What fun!

An Exciting Discovery on the VTOL Test Stand

Our flight tests of theYH-41 helicopter with its variable stability capability and the presence of the VTOL Test Stand, which needed a research program to prove its value, presented a serendipitous opportunity: Why not mount the YH-41 on the test stand, exercise its variable stability capability, and actually measure the forces generated? A relatively cheap research program. Sounded great!

But… the variable stability program incorporated a thing called a "rate gyro." This gyro would help stabilize the helicopter when the aircraft developed roll rates — On the static test stand, it could not generate any roll rates since it was bolted firmly to the test stand.

Undeterred, we proceeded anyway, because we still needed to prove the test stand and we could still generate vertical, roll, yaw, and longitudinal control forces on the helicopter and measure them.

We had a simple mounting cradle built, mounted the YH-41 on the cradle, Gerry Tebbens was assigned as "flight" test engineer, and Bob Baldwin "flew" the aircraft. The tests were successful, and we proved the test stand and gathered some interesting (albeit useless) data on the YH-41—useless because Cessna built no more of the helicopters and they were retired from service.

Cessna YH-41 on VTOL Test Stand

The most valuable product of these tests, however, was a paper on the VTOL Test Stand that Bob Baldwin and I co-authored, and that Bob presented to the Society of Experimental Test Pilots (SETP) in Los Angeles that year.

Now, as you can imagine, a test pilot symposium is a bit more—shall we say—relaxed than an engineering symposium. That is not to say that the test pilots don't take their jobs just as seriously—they just like to have fun as well.

So, with this in mind, we planned to end the short film clip of the YH-41 tests on the VTOL Test Stand with a smashing conclusion.

We asked one of the (better looking and more adventurous) secretaries if she would pose on the test stand in a bikini bathing suit, and she agreed. We set up the photo shoot.

We had no trouble getting photo support for this test! There wasn't a single cameraman with lower rank than Master Sergeant, and there was plenty of Photo Department supervision!

After Bob presented his written paper, he introduced the film clip of the YH-41 tests. The film showed the various tests being conducted.

Then it cut from showing the helicopter running on the top of the elevated test stand to Gerry Tebben in the control room. It showed the recording needles going wild, and Gerry's headphones flapping. Gerry pushed the button to lower the test stand and ran to the window to see what was happening.

The film showed the hydraulic rams as they came down and the top of the test stand came slowly into view:

The helicopter was *gone*!

In its place was this pretty girl in a bikini, who smiled sweetly, and stretched…

Our paper was the hit of that year's SETP symposium!

These tests also proofed the VTOL Test Stand, and it was used for several VTOL and helicopter tests after that.

An added benefit: I got to see Gordon Cooper again, one of the NASA Mercury 7 astronauts, with whom I had worked on my very first flight test program back in 1957.

XV-5A on VTOL Test Stand

Bell* Huey *on VTOL Test Stand

An Offer That I Could Refuse

In 1961 I was responsible at Edwards for monitoring all of the VSTOL (non-helicopter) flight research in the world. I collected data on the U.S. Air Force, Army, Navy, NASA, and industry programs, and the Canadian, British, French, German, and Russian STOL and VSTOL programs. I learned as much as I could about their design concepts, research status, and test results.

I was also assigned to prepare for the forthcoming Tri-Service VSTOL test programs scheduled to come to Edwards. This involved both the Ling-Temco-Vought XC-142A tilt-wing VSTOL transport and the Curtiss-Wright X-19 tandem-wing tilt propeller VTOL aircraft that were jointly funded by the Air Force, Army, and Navy–hence the Tri Service designation. I also had to monitor the Navy/Bell Aerospace X-22 tandem wing tilt-duct VSTOL aircraft program that was being managed and conducted by the Navy Flight Test Center at Patuxent River Naval Air Station in Maryland.

Specifically, I had to write the Air Force, Army, and Navy Category II performance, and stability and control test plans, for the XC-142A and X-19, that would satisfy all three services who had invested heavily in these development programs. I had to define the special flight test instrumentation that would provide the data needed to document the flight characteristics of these unique aircraft that sometimes flew like an airplane, sometimes like a helicopter, and sometime like neither one–what kinds of special instrumentation would provide the flight test data I needed to define and document the aircrafts' unusual flight characteristics and what would we do with these data once we had them? I had to define the mathematical equations that would enable us to document the flight test results that would be meaningful for an accurate assessment of the aircraft's mission capabilities. Fortunately I had Edwards flight test engineering help from experienced professionals like Clen Hendrickson, Bob Reschak, Gerry Tebbens, and Burt Rutan (yes, Burt Rutan), to work the data reduction equations.

I also had to set up the flight test organization at Edwards that would satisfy all three services that their interests were satisfactorily represented. I had to work closely with Edwards' supply and maintenance organizations to set up the resources to maintain the aircraft, including hanger and office space. In addition, for the several years until the arrival of the rest of the Tri-Service flight test team, except for the engineering support, I *was* the Tri-Service VSTOL Test Force. To keep track of everything that was going on (or needed attention) I scheduled monthly meetings with everyone at Edwards involved with these programs. Surprisingly, some of the most meaningful results came from meetings at which we thought we didn't need a meeting, but discovered serious problems that we resolved before they impacted the program.

This was great fun, but I was concerned, since I had been at Edwards since 1957, that I was due for transfer to a desk job at Wright-Patterson AFB in Ohio. I didn't want this because I thoroughly enjoyed my "hands-on" test work at Edwards. Why fly a desk when you can fly real aircraft?

I knew that if I was appointed to post graduate school by the Air Force that I would be left at Edwards until my schooling started. With this in mind, I went to the base Personnel Department and inquired about my likelihood of transfer.

Captain Martin took out my file and confirmed my suspicions: "You've been here 5 years. You are very ripe for transfer!"

"I like what I'm doing here, and want to stay awhile longer. I understand that if I apply for graduate school, then I would stay here until I am approved and school starts." Subconsciously I was thinking: [*"That gives me plenty of time to figure out what to do next!"*].

"That's *almost* right," Captain Martin corrected me, "but not until you are actually *approved* for graduate school. And, your chances of being approved would be greatly improved if you had a *regular* Air Force commission instead of just your ROTC *reserve* commission."

"OK. I'll apply for a regular commission," I said, figuring that *that* would take a long time because they were usually difficult to get.

I returned to my desk, and so help me, there was a letter there waiting for me that read, in effect: "Congratulations! You have been approved for a Regular Air Force commission. Just sign here and pass the physical examination!

WOW!

I was very proud of my record at Edwards–especially so because of the incredibly high caliber of the fine professionals I was working with. Edwards was the premier flight test center of the world. Thomas Wolfe, in his 1979 epic story of NASA's Space Program, *The Right Stuff*, talked of the Navy flight test center at Patuxent River, Maryland, and the British and French flight test centers, but he referred to Edwards as "the pyramid of flight test, with Chuck Yeager firmly at its apex."

At my very first performance review my supervisor told me that my high performance was typical of the average Lieutenant Colonel's—not bad for a brand new 2nd lieutenant! I had been promoted to 1st Lieutenant after only six months active duty. My idea of asking the Strategic Air Command tanker to wash our B-58 windshield of hydraulic fluid, devised at the same time as the pilot Charlie Bock, enabled us to land the plane safely at Edwards and complete our mission the next day. By replacing the 95,000 pounds of highly flammable jet fuel with water in the B-58 refused takeoff tests I probably saved the airplane when the high level fuel shutoff valve failed and the water–not fuel–cascaded over the wing onto the blazing wheels. By conducting the dangerous short takeoff tests, before the engine damaging landings, of the Boundary Layer Control NC-130 BLC during Project Rough Road Alpha takeoff and landing tests in the soft sand at Yuma Naval Air Station it may have saved the airplane and crew when an engine failed during takeoff run up after the tests were safely completed. By determining the natural frequencies of the VTOL Test Stand I may have prevented a serious accident at the critical wheel height. By advising Bob Baldwin that we should land immediately after a piece of the rotor broke on the YH-41 helicopter, it may have saved both us and the chopper.

For this work I was recommended by my superiors and honored by my technical society, the American Institute of Aeronautics and Astronautics, by being promoted to the rank of Associate fellow. Article III 3.2 AIAA Constitution, specifies that "Associate Fellows shall be persons who have accomplished or been in charge of important engineering or scientific work, or who have done work of outstanding merit or have otherwise made outstanding contributions to the arts, sciences, or technology of aeronautics or astronautics." Furthermore, only one AIAA member is permitted advancement to

Associate Fellow grade for each 150 AIAA members—that's only seven tenths of one percent. Quite an honor, especially at The Air Force Flight Test Center!

I never expected special recognition for these accomplishments–they were just part of my job, routine, what I was paid to do, and others were doing just as incredible things. But probably it was my three outstanding Officer Effectiveness Reports (OERs) that had been noticed by the Air Force personnel department!

I telephoned the personnel officer immediately, and told him: "WOW! You guys sure work fast! I'll let you know."

My wife Paula and I discussed this. It was obvious that I was on a fast track for higher rank if I stayed in the Air Force, but that would certainly involve leaving Edwards–probably for a desk job at Wright-Patterson AFB pushing paper instead of flying exciting airplanes. I could confidently look forward to retiring a full colonel, but it was very rare that a non-rated (i.e., non-pilot) would be promoted to general. The retirement benefits would be good, but I could certainly make a better income on the outside, although money was not the primary issue. I knew of many retiring officers who had no equity built up in home ownership, and had to essentially start from scratch when they retired and left free Government housing.

The end assessment was that we would probably be better off in civilian life than in a blue-suit Air Force career, and I could stay at Edwards as long as I wanted. Once we decided to not accept the regular commission, we also realized that we had made the decision to get out of the Air Force as soon as possible.

I advised the personnel officer of this decision, and signed the necessary papers to be released from active duty (I would still have a few years of Reserve duty obligated from my ROTC commission). It would be a year before I could get out of active duty, because of the Vietnam War, but, in due course, we would be civilians.

I made Captain in October, 1962, after only five years of active duty. I seemed to be on a "fast track" for rapid advancement, but the original decision to resign from active duty still seemed valid. My career options seemed better in civilian than in military life. I

Capt. Rob Ransone, Shortly Before Retiring From Active Duty to Become a Civilian GS-13

believed that I could never find another Air Force job as exciting, interesting, and rewarding as my work at Edwards.

My "blue suit" experience qualified me for a GS-12 level with civil service, but I asked my boss, Major Lou Setter, if he couldn't do better than that. He asked me to tell him what job title I could think of for the forthcoming Tri-Service VSTOL program that would justify the higher rank. I thought a moment before answering:

"This is a Tri-Service organization, with Tri-Service funding and with Air Force, Army, and Navy flight test engineers under my supervision. I need to give the Army and Navy engineers appropriate recognition. For the XC-142A I want to assign the lead Army engineer as my deputy, Clen Hendrickson as Project Engineer for the Performance flight tests, and the lead Navy engineer as Project Engineer for Stability and Control flight tests. For the X-19 I want to assign Bob Reschak as the Project Engineer with the Army and Navy engineers as joint project engineers. Therefore my title should probably be Chief Flight Test Engineer."

He agreed, and I passed the qualifications review and would continue the same job, but now as a GS-13 civilian. I signed out of the Air Force on a Friday afternoon, signed on in civil service the following Tuesday morning, so I just worked a half-day on Monday—why not? I had to do it, anyway, and I was having so much fun!

This was a major decision, not only for my career, but also for our entire family. A major decision that Paula and I made together. Was it the right decision? I believe so: it was the best decision at the time, and lead to a completely different life than we would have lived had I stayed in the Air Force.

Years later, as a Visiting Associate Professor at the University of Virginia, I would teach my graduate students that there are four kinds of decisions: a good decision, a bad decision, the right decision and the wrong decision. A good decision is one that is made deliberately based upon all the information available at the time, with full awareness of the expected consequences. A bad decision is made flippantly, ignoring some factors or with little regard for their consequences. A good or bad decision may turn out eventually to be either the right or wrong decision—no one can predict the future. The proof of a good decision is demonstrated through the test of time, in that, regardless of whether it was right or wrong, would you make the same decision again, based solely upon what you knew at the time?

I believe that throughout life we make thousands of decisions, sometimes with great care and deliberation, sometimes spontaneously on the spur of the moment, and sometimes out of immediate necessity, but each of these decisions can have incredible consequences that affect not only our own lives, but untold lives around us. For example, an extra cup of coffee in the morning can lead to a time-sequence of events that can either result in a fatal car crash at an intersection or a safe trip to work. What would my life, and the lives of my family have been had I not decided to work at General Dynamics after graduation and learned about the Air Force Flight Test Center? What if I had gone to Hamilton Air Force Base instead of Edwards—I would never have met Paula and had our two wonderful children? What completely different chain of events would have followed had Colonel Lane not happened through Base Operations that day and advised me not to fly on the B-57 that crashed on landing. Who knows what repercussions would have followed if we had not made that quick decision to wash our windshield in the B-58 that day over Edwards. If I had left Edwards I would not later have been planning the flight tests of the XC-142A–a special test I planned required that one of our test pilots be with us instead of on the XC-142A in Dallas that crashed, killing the entire flight test crew. If I had not moved to the East coast after leaving Edwards in 1968, I would not have been traveling between New York and Texas in time to dash behind a car to snatch a young child out of the way before a driver backed over her. Nor would I have been in a filling station in Dallas one day in time to grab another small child off the back bumper of a station wagon that was driving off, with the driver oblivious that the child had climbed out of the back window and was clinging to the back of the car.

In Chaos Theory, there is the concept of the Butterfly Effect: the concept is this, that the smallest events, even as small as where a butterfly alights, can have enormous future consequences.

Based upon my criteria of good or bad, right or wrong decisions, in retrospect, with full advantage of 20/20 hind sight, I fully believe that our decision to get out of the Air Force was a good decision, as well as the right decision. What if I had accepted the regular Air Force commission and followed an Air Force career? I might have helped develop some fantastic new aircraft or done something that saved a hundred lives? But that is pure speculation—what I actually did was real.

There is a philosophical theory that every eventuality actually happens, but we are aware of only one. In another Space-Time Continuum I may have taken a different path, and, who knows? Maybe that path actually exists somewhere?

Each decision directs us on a specific branch of life. At each fork in the road, once chosen, we can never change our minds or go back Learn from past triumphs and mistakes but accept whatever the consequences and harbor no regrets. This key decision led our family on a completely different road than if I had followed an Air Force career.

Except for the fact that we moved to a smaller house, for which we paid a small rent, and could no longer shop at the base Commissary for food or the Base Exchange for personal items, our life was much the same. We still belonged to the Officer's Club, and still had the same friends and entertainment. I was still doing the same job every day.

And thus, as one wonderful chapter of our life ended, we eagerly anticipated the next.

The 11th century Persian poet, mathematician, and astronomer Omar Khayyam, in his rambling collection of quatrains: *The Rubaiyat*, wrote:

> The moving finger writes,
>
> and having writ, moves on,
>
> nor all your piety nor wit can
>
> lure it back to cancel half a line,
>
> nor all your tears wash out a
>
> word of it.

My VSTOL Briefing to General Branch

In 1961, General Electric won a US Army contract to develop its fan-in-wing VSTOL concept, the XV-5A, with design, construction, and flight testing of the aircraft sub-contracted to Ryan. General Electric retained responsibility for the propulsion system and integration with the aircraft. In the inboard portion of each wing a 5-foot diameter turbofan provided vertical lift. A smaller fan in the nose in front of the two person cockpit give pitch control and additional lift. These fans were driven by exhaust gas from two GE J85 turbojet engines through turbines on their fan blade tips. Retractable covers enclosed these fans for forward flight, which was provided by the normal jet exhaust through the tail.

Army/General Electric/Ryan Fan-In-Wing VSTOL Aircraft Hovering

Ryan wanted to test the experimental aircraft at Edwards because of the safety of the dry lakebed and Edwards' isolation from populated areas. General Branch, the flight test center commanding general, had to make a decision as to whether he should commit center resources (mostly hanger space and crash recovery) to support this program. Colonel Guy Townsend, Flight Test Division commanding officer, set up a meeting for me to brief General Branch on the status of forthcoming DoD VSTOL programs to show why he should support the XV-5A VSTOL aircraft program.

General Branch was a really, nice, guy! He was very laid back. In fact, his nickname was "Twig." Can you imagine a Brigadier General whose name is "Twig" Branch? He was tall and husky. He smiled a lot and was soft spoken but very competent and knew what information he needed in order to make his decisions. His management style was to work *with* his officers–all of them–not to lord it over them. This was exemplified by his request for me to brief him on the VSTOL programs instead of having one of his senior officers relay the information to him second hand.

Colonel Townsend was something else! He was a Texas Aggie. He enjoyed "hazing" the other officers by kidding with them, except they didn't always know when he was kidding. I was reminded of his personality years later in the movie *Patton* when George C. Scott, as general George S. Patton–a notoriously volatile man–had had a lively interchange with his staff one day. After leaving the briefing, his aide told him: "Sir, sometimes they don't know when you are joking." George replied: "It doesn't matter whether *they* know, it only matters that *I* know!"

The "hazing" at A&M consisted mostly of upper classmen chastising lower classmen to train them in the ability to take strong criticism without losing their focus on the job at hand and calmly correcting any problems. Don't take it personally. At A&M, the sophomores hazed the freshmen, the juniors hazed the sophomores and freshmen, and the seniors hazed everyone! Since I realized what he was up to, and he was confident that I knew what I was doing, we got along great!

Colonel Townsend was the only full colonel on the base that the other full colonels called "Sir!" He enjoyed this immensely. Enlisted airmen, driving to the Los Angeles airport to pick up returning officers, lived in dread of Colonel Townsend. A tall, slim man, he had a slightly hooked nose, piercing, deep set eyes, and a way of looking at you that reminded you of the pictures of bald eagles glaring at you from patriotic posters. And the silver eagles on his shoulder epaulet, warning others of his rank, enhanced this persona.

When he was addressing someone and wanted to know whether they were empowered to speak and make decisions, he would ask: "Are you the HMFIC?" When asked what that meant, he would answer (sorry for the use of the "F" word): "That means the Head Mother Fucker In Charge." I asked him one day if he was the HMFIC of Flight Test, and he replied with a twinkle in his eye: "No, I'm the HMFIC*C* – I'm in *complete* charge!" And, indeed he was. The thing that enabled him to continue these antics was that he was incredibly capable, and after leaving Edwards he was promoted to two-star general and headed up the USAF B-1 bomber development program at Wright-Patterson Air Force base.

An example of Col. Townsend's impatience with being mislead was his handling of Lockheed Aircraft Corporation's spin demonstrations of their F-104 *Starfighter*. Basically, Lockheed didn't want to spin the aircraft. The F-104 had a major longitudinal stability and control problem. The wings were very thin for low aerodynamic drag at twice the speed of sound, and had a short wingspan. The wing thickness ratio was the same as a Gillette double edged razor blade! The aircraft was known as "The Missile With A Man In It." The fuselage was long and slender, but the wide air inlets for its powerful J-79 afterburning engine (the same engine as those in the B-58) along side the fuselage at the wing roots created considerable lift. Normally, this was OK. But if the aircraft stalled, and the wings quit flying, instead of the nose dropping down as with most airplane, the lift from these engine inlets lifted the nose, and the wing's "shadow" blanketed the horizontal tail, which stalled the horizontal tail, and the airplane did a back flip over its own shoulder! This was called a "pitch up." Not good, not good at all!

Lockheed studied dozens of horizontal tail configurations, including twin tails, biplane tails, high tails, low tails, mid tails, horizontal ails with vertical end plates. You name it—Lockheed tried them all, but with no luck. The pitch up persisted. The T-tail (with the horizontal tail at the top of the vertical tail) was the best solution because, even though the aircraft would still pitch up, the pitch up did not occur within the normal flight regime. Lockheed's solution was to incorporate a "stick pusher" that shoved the pilot's control stick forward if he was getting too close to the pitch up

***Lockheed F-104* Starfighter**

region. In the landing configuration, however, the stick pusher was disengaged and a stick shaker warned the pilot.

During one F-104 flight test demonstrating the pitch up, Air Force test pilot Bob Rushworth was chasing. After the flight, Bob remarked: "I don't know what that last maneuver was, but you sure lost me!" Bob reported that the F-104 flipped up 90 degrees to the vertical, immediately flipped 180 degrees nose down, then flipped 360 degrees up over its own shoulder flying backwards.

To spin the aircraft, Lockheed had to enter the spin through a pitch up, but that was only part of the problem. The Northrop F-101 *Voodo*o had a similar problem; it would pitch up, and then go into a flat spin that was not recoverable—the crew had to eject.

Lockheed didn't want to do this because they expected the F-104 to enter an unrecoverable flat spin, and were about halfway through their "snow-job" of explaining to Colonel Townsend why the spin tests were not really necessary, when Colonel Townsend suddenly exploded, told them what he thought of their feeble excuses, told them to get their %@$%^#*&^@# act together, and stormed out of the room!

After a very quick consultation among themselves, the Lockheed folks agreed to the spin demonstration tests provided they intsalled a "spin chute" at the F-104 tail to recover in the event the airplane could not be recovered aerodynamically. The spin tests were subsequently completed safely, and proved that the airplane could be recovered from the spin if it had enough altitude. The airplane would pitch up, flip over backwards, tumble, enter a flat spin, roll over into an inverted flat spin, the nose would drop, and the airplane recovered.

Colonel Townsend respected my ability, but he still wanted to be there when I briefed his boss, General Branch, on our VSTOL programs because it was a very important issue.

I knew that the higher up a person is, the less time and patience he has for briefings. His time and attention are very valuable. I also knew from my Toastmasters training that I needed to be specific and to the point, and have a good summary at the end of what I wanted him to remember. I had prepared a half dozen flip charts with the needed information.

I had conducted a little experiment one night at Toastmasters when I gave the Educational Tidbit. I had compiled some statistics on Southern California drivers, and presented this detailed information in my presentation. Without telling my audience, I had divided my presentation into three parts. During the first part I simply stated the statistics: "There are 6 million licensed drivers in Southern California." In the second part, I emphasized the statistics: "In Southern California, 2 million of the drivers are either handicapped, or under the influence of drugs or alcohol–2 million! 2 million drivers!" For the last part I showed them a chart with detailed statistics tabulated. I could see the puzzled expressions on their faces: "What has *this* got to do with Toastmasters?"

After I finished my speech, I took down the chart and handed out a "pop quiz" that required them to answer specific questions on the statistics. I scored the answers and presented the results at the end of the meeting. Where I had only briefly mentioned the statistics, only 26% of the answers were correct. Where I had emphasized the statistics, 34% of the answers were correct. Where I had showed them a picture of the information, an incredible 86% of the answers were correct! I have since seen these same results from similar experiments.

This is how I knew that I must have a clear, concise summary chart at the end of my presentation if I wanted to get my point across.

North American X-15 Rocket Research Plane

The afternoon arrived, and Colonel Townsend and I waited in General Branch's office. I was supposed to have an hour, but he was already a half hour late.

In addition to that, Captain Bob White was flying the X-15 to Mach 6–

about 4,450 mph–at the same time, and General Branch kept running to the window to see if he could see Bob's contrails.

And in addition to *that*, an Air Force sergeant had gotten run over and killed at a railroad crossing the night before, with another man's wife, and the sergeant's mother telephoned three times during my presentation! Of course, General Branch had to talk with her.

But, even with all of these incredible distractions, I was prepared. When I got to the end, I looked General Branch right in the eye, and said: "General, if I had only two minutes of your time, this is the only chart that I would show you!"

"Oh, really?" He adjusted his glasses, and I had his undivided attention for two minutes while I showed him the summary chart. This made such an impression that he agreed to support the Army's XV-5A program, and repeated the exact information in a speech to the Lancaster Chamber of Commerce two weeks later.

The eagle smiled.

An Amazing Trip

Paula's lifelong friend and high school/college classmate, Carol Ann Bright was getting married in Waco, Texas, so of course Paula attended her wedding as Matron of Honor. Since I couldn't take off from my responsibilities preparing for the upcoming XC-142A flight tests at Edwards, I put Paula and two-year old son Key on a plane in Los Angeles, and went back to work.

Then fate's fickle finger beckoned me, and I began one of the strangest trips of my career.

There was to be a formal review of the XC-142A test program at LTV in Dallas, so I packed my bag, got a travel advance and tickets, and flew to Fort Worth. On such trips I stayed in a local hotel with the other attendees, but LTV would generously lend me a company car to visit Mother in Fort Worth for dinner.

After the meeting, I discovered that I needed to attend a Tri-Service Program Review of the Curtiss-Wright X-19 at the System Program Office at Wright-Patterson AFB, in Dayton, Ohio. The travel office at LTV was able to arrange my plane tickets, and I flew to Ohio.

After the meeting, Captain Bob Baldwin, Chief USAF Test Pilot for the X-19, and I decided that we really needed to see what was happening at Curtiss-Wright Aircraft Company in Caldwell, New Jersey, (about 30 miles from New York City). I got more tickets and another travel advance, and flew to New York with Bob.

This was my first trip to New York City, and I was excited when the pilot announced: "Please raise your tray tables and fasten your seatbelts for landing at New York's Idlewilde Airport." This was before the airport was renamed for President Kennedy. The lights of the city were awesome as we made our approach and landed. Bob knew of a hotel with rates acceptable to our travel allowance, and we enjoyed strolling around Times Square that night. He also knew of a little Times Square cafeteria where we could get a small steak, baked potato, and green salad for $1.85. Yes: that's one dollar and eighty-five cents. Remember, this was 1962.

The next morning, Curtiss-Wright sent a car to pick us up and I got my first real taste of dodging through New York City traffic to the Lincoln Tunnel, and west on US 46 to Caldwell, New Jersey. We finished our examination of Curtiss-Wright's progress on the X-19 preparations, and decided what to do next. Bob wanted to fly the variable stability helicopter that NASA Langley Flight Research Center was flying in Hampton, Virginia, but it would be a few days before that could be arranged.

My sister, Sally, was dancing in the musical comedy *Top Banana,* starring Milton Berle, in Owings Mill, Pennsylvania, so Bob and I caught a train from New York's Penn Station to Philadelphia, and then a cab to Owings Mill.

By the time we got there, the show had started, but Sally had left us tickets at the box office, so we got to see most of it. Show people usually don't eat dinner before a performance, so we went to dinner after the show with Sally and some members of the cast. "Uncle Milty" was too important to socialize with us peasants! I introduced Bob and when the cast found out that he was an experimental test pilot they were very excited and wanted to know all about his adventures. After a bit of ego satisfying, Bob announced that I was also an experimental flight test engineer and flew with him. Then the cast turned their attentions to me, and I got my share of the ego boost.

What fun.

The next morning, Bob and I caught a bus to Philadelphia and a flight to Newport News, Virginia. Bob's friend, Captain Paul Currie, the Army test pilot on the variable stability helicopter, picked us up at the airport, and took us to NASA Langley to see the aircraft. Bob and I had developed a glossary of STOL, VTOL, and VSTOL definitions, and NASA was developing handling quality criteria for these unique aircraft. NASA Langley's variable stability helicopter was a key resource in those studies. The helicopter had a special on-board computer that could control the helicopter's flight characteristics and the effects of the pilot's control inputs to the helicopter's responses. They could then simulate various VSTOL aircraft flight characteristics, derived from computer calculations and wind tunnel tests, and determine the aircraft controls needed for safe flight.

After we finished a very interesting test flight, Paul flew us to the airport, where we boarded a commercial flight back to Wright-Patterson AFB. After a brief meeting at the Systems Program office,

Bob flew back to Edwards, but I found that I was needed back at LTV to define the special instrumentation needed for our Category II performance and stability & control flight tests. I got additional tickets, another travel advance, and flew back to Fort Worth.

After my meeting at LTV, I borrowed Mother's car, and drove to Waco, arriving just in time for Carol Ann's wedding!

The next morning, Paula, Key, and I drove back to Fort Worth, spent a day with Mother, and then we flew back to California together.

I had been gone about three weeks, and when we got the car from the LAX parking lot, I didn't even bother to drive up to the ticket window—I just pulled over to one side, walked up to the window and offered my BankAmericard to the cashier.

An amazing trip that I couldn't have planned in advance!

Although this trip was the most complicated I ever took at Edwards, there was another that was unusual in another way: Trying to fly home from Wright-Patterson AFB during a summer thunder storm.

My flight home, departing the Dayton airport one summer afternoon, was delayed because of severe thunderstorms in the area. The local thunderstorm was so severe, in fact, that our plane from Columbus, Ohio, couldn't even land at Dayton. After circling for 45 minutes or so, it returned to Columbus. After an hour and a half, it returned, circled a half hour or so and finally landed.

We boarded the plane, but sat on the runway for a half hour or so waiting for the thunderstorms to clear. After the storms persisted, we taxied back to the terminal and deplaned. After another hour's wait, we boarded the plane and managed to takeoff for our short flight to Chicago, where I would change planes for Los Angeles.

After circling Chicago for a half hour, we returned to Dayton, where we once again deplaned and waited. Finally, after boarding the plane and a rough flight to Chicago, we started our descent. It was very, very rough, and suddenly there was a sharp crack sound and a bright blue light flash just outside my window.

"What was that?" My seatmate demanded, very frightened!

“Just some lightning outside,” I answered. I knew very well that our plane had been struck by lightning, but I didn’t want to alarm her any more. I also knew that airplanes are frequently struck by lightning during thunderstorms aloft. Normally the huge electrical charge passes harmlessly from one wing tip to the other, existing the highly conductive aluminum structure with no damage. Since we passengers were not grounded, and aluminum is a much better conductor of electricity than our bodies, we felt nothing.

We finally landed at Chicago about three o’clock in the morning. My connecting flight to Los Angeles had not even landed at Chicago. The airlines put us up at a local hotel for the night, and I continued my flight home the next morning.

Incredibly, as I was walking through the terminal at three o’clock that morning, half asleep, and completely worn out, Bob Nagle, another engineer from Edwards, came walking by, and nonchalantly announced: “I’ve got a message for you from Edwards.”

What are the odds?

Federal Republic of Germany VSTOL Programs Assessment

One of my first jobs as a civilian was a trip to Germany.

In 1963, the U.S. Department of Defense (DoD) was cooperating with the Federal Republic of Germany (FRG) to develop VSTOL aircraft. The DoD had invited the Germans to the U.S. to see our development activities, and the Germans reciprocated. So , in May of 1964, I found myself on my way to Germany as the flight test representative of a 13-person DoD team to assess FRG's ability to develop VSTOL aircraft. As guests of the German government, we were treated royally!

Our team flew to Frankfurt and rode the Schnell (high speed train) to meet at the American Embassy in Bonn to make our plans. Our first tour was to Friedrichshaften, in the southwestern part of Germany just across Lake Constance (Bodensee) from Switzerland, to assess Dornier's DO.31 VSTOL jet transport program. Our flight was in a Luftwaffe transport, with the German Iron Cross logo on the wings and fuselage. I'll have to admit to a few moments of uneasiness, looking out at those insignia on the wings, remembering the past, but, of course, there were no signs of the notorious Nazi swastikas, and our hosts were most hospitable.

Flight to Friedrichshaften in a Luftwaffe airplane

George Bright, because of his experience with the X-18 VSTOL transport program, had taken a job with the German aircraft company EWR to help them design their experimental VJ-101 six-jet VSTOL fighter. When Dornier started its design of the DO.31, he recommended another American test pilot, Drury Wood, who took a job similar to George's, at Dornier. George still had a strong hand in the DO.31 design.

The Dornier Do.31 was developed to West German requirements as a prototype VSTOL transport able to support combat aircraft in the field. Do.31 was powered by two British Rolls-Royce Pegasus 5 v-t engines equipped with diverter vanes (the same as the British P.1127 and Harrier), one under each wing. The jet exhausts, and part of the compressor exhausts, could be deflected downward to provide upward thrust for vertical takeoff and landing, or aft to provide forward thrust for cruise flight. It also had eight Rolls-Royce RB.162 lift engines, four in each wingtip pod. With ten Rolls-Royce engines, Rolls-Royce *loved* the Do.31. Although it performed its VSTOL flights successfully, in the final assessment it was decided that it had way too many engines to be economically feasible. Ideally, an efficient VSTOL aircraft should minimize the weight and space of anything that is used solely for takeoff or landing.

Federal Republic of Germany/Dornier DO.31 VSTOL Transport

We were wined and dined royally, and, in addition to examining the DO.31 (which we were not able to see fly), we toured the model exhibits of all of Dornier's fascinating airplanes. These planes, from the company's first beginning, were extraordinary. Claude Dornier was a Frenchman and had spent the war years in Switzerland. This had not pleased the Nazi regime, but now all was forgiven.

We were treated to a wonderful evening at a local beer hall. The buxom serving girls, charming in their white frilly Bavarian blouses and leather bustiers and long skirts, carried five one-liter steins of beer in each hand. There was wonderful entertainment that included a young man and woman from Switzerland who yodeled. I was used to the American western yodelers, with their "ye olde lady hoooooooos" that I disliked a lot! But this was different: With their two soft voices yodeling in bright harmony they sounded like tiny silver bells. It was beautiful!

Our next stop was at Entwicklungsring Süd (EWR), in Munich (Bavaria), where we examined George Bright's VJ-101, and talked to him and the VJ-101 design and flight engineers. This single-seat experimental VSTOL aircraft had six engines: two were located at each wingtip and could be swiveled from the horizontal position for cruise flight to the vertical position for vertical takeoff and landing. As they were rotated from the vertical to the horizontal, the aircraft would accelerate and transition rapidly from the vertical to the forward fight mode. Two additional engines were mounted

Federal Republic of Germany EWR VJ-101 VSTOL Fighter

vertically in the fuselage directly behind the pilot and were used only for vertical or very short takeoff and landing. You can see the engine inlet doors open behind the cockpit in the photograph. We did not see this aircraft fly.

Some of these engineers were not too happy with George, and told me several stories that came in useful a few years later when the US and FRG had an agreement to jointly develop a VSTOL fighter prototype.

Basically, the problems were the flight control system (which I will not go into now), the fact that George had not allowed anyone but himself to fly the aircraft, and that there were no formal preflight or landing procedures. They were all in George's head. The later had resulted in George taking off vertically with the parking brake set (that's OK), but then forgetting it was set and making a rolling landing at 200 knots with the parking brake set (that's *not* OK). Blew a few tires, but not fatal, either, fortunately.

One thing that I will mention about the flight control system: it was very automated, and the aircraft was *unflyable* if the automatic control didn't work properly. George had designed a "triplicated" system so that if any one of the three systems disagreed with the other two, the two in agreement would lock the first one out of the circuit.

"But what if the system fails?" I asked the chief engineer.

"*Fail*?' He echoed, aghast. "It is not designed to *fail*, it is designed to vurk! It vill vurk!"

Finally, I found someone who had at least heard of "mean time before failure (MTBF)." But he had an answer, also: "If the mean time to fail is 100 hours, ve vill change it at 90 hours. It vill vurk!"

I could not convince him that 100 hours MTBF didn't mean that it *wouldn't* fail until 100 hours but that in the aggregate, the engine fleet could be expected to run, on average, 100 hours without failing. An individual engine could, in fact, fail the first hour.

Skipping ahead a year: On a rolling takeoff one day, as the aircraft lifted off the runway and the stabilization system came on automatically, the system decided that the aircraft was rolling to the right, and commanded full left roll! Someone had installed one of the attitude gyros upside down! Suddenly George was staring up at the runway—not good! He counted to " ½, " as he relates it, and pulled the ejection seat handles. He was ejected horizontally out across the grass as the aircraft continued its roll and subsequent fiery crash. I have movies of this: The photographer, filming the takeoff, started running away when the aircraft started to roll, occasionally looking over his shoulder to take a few more feet of movie, without ever taking his finger off of the camera record button—aircraft lifting off, starting to snap roll, sky and ground, sky and ground, George lying on the ground, sky and ground, sky and ground, VJ-101 burning on the runway, sky and ground, sky and ground!

Fortunately, George survived, and continued the flight tests after the aircraft was rebuilt. With six engines, this concept also proved too expensive to operate, and never went into production.

From Munich, our German hosts took us to visit Berchtesgaden, Adolph Hitler's notorious *Eagles Nest*, in Obersalzberg, high in the Bavarian Alps. It was a beautiful, clear day, and the view of the Bavarian countryside, thousands of feet below, was breathtaking. It was here, in his bunker, that Adolph Hitler committed suicide with his mistress, Eva Braun, in the final days of the war, and his guards burned the bunker to destroy the bodies. On this quiet, peaceful day with our attentive hosts, there was no hint of those horrors long ago.

***Berchtesgaden, Adolph Hitler's* Eagles Nest**

The bunker where Adolph Hitler and Eva Braun committed suicide

Our last visit was to review the early development of the BAK-111 jet fighter program at Vereinigte Flugtechnische Werke (VFW) in Bremen, north Germany. VFW was a consortium consisting of Focke-Wulf, Heinkel and Weser. This design eventually became the VAK 191B, an experimental German VTOL nuclear strike fighter. VAK means Vertikalstartendes Aufklärungs und Kampfflugzeug (V/STOL Reconnaissance and Strike Aircraft.) (Aren't you sorry you asked?) The photograph shows the VAK191B under the wing of the DO.31, whose engine you see overshadowing the VAK-191B.

Federal Republic of Germany Vereinigte Flugtechnische Werke (VFW) VAK 191B VTOL Nuclear Strike Fighter

One of the most interesting events during this trip to Bremen, was our dinner in the Rathskeller. The building was 100 years old when Columbus sailed for the new world. It also had a fantastic wine list. Normally the cellar stocked 2 ½ million bottles of wine. The oldest was a 1735 white wine that I wish I had bought ($12.00) just to put in my wine rack. The most expensive was a 1938 *Trockenbeerenauslese*, an incredible sweet desert wine similar to the wonderful French *Château d'Yquem*. A quick explanation of these two great wines, but first, a description of the Bordeaux wine classifications

The Bordeaux Wine Official Classification of 1855 resulted from the 1855 Exposition Universelle de Paris, when Emperor Napoleon III requested a classification system for France's best Bordeaux wines which were to be on display for visitors from around the world. Brokers from the wine industry ranked the wines according to a château's reputation and trading price, which at that time was directly related to quality. This classification remains widely used today. The brokers classified four Chateaux as First Growths (Premiers Crus): *Château Lafite Rothschild*, *Château Latour*, *Château Margaux*, and *Château Haut-Brion*. *Château Mouton Rothschild*, because its prices were a bit below the others, but still higher than the Second Growth (Seconds Crus) wines, was classified a Second Growth. Baron Rothschild didn't like this, and he and his heirs lobbied the French government to change its classification, and finally, after 98 years, in 1953, *Château Mouton Rothschild* was reclassified a Premier Crus.

But there was one wine that so far outclassed the rest, that it was classified a Superior First Growth (Premier Cru Supérieur): and that wine, a sweet desert, white Sauternes, is *Château d'Yquem*. In the Bordeaux Wine Official Classification of 1855, *Château d'Yquem* was the only Bordeaux given this rating, indicating its perceived superiority and higher prices over all other Bordeaux wines. Wines from Château d'Yquem are characterized by their complexity, concentration, and sweetness. In a good year, a bottle will only begin to show its qualities after a decade or two of cellaring and with proper care, will keep for a century or more, gradually adding layers of taste and hitherto undetected fruity overtones. Château d'Yquem is also purportedly the only house that ensures that the wines are made from individually picked grapes, ostensibly to ensure that quality is the finest possible. This means that each grape vine, on average, can only produce a single glass of wine. *Château d'Yquem* is made from grapes on which a special mold has developed, known as the *noble rot.* Although the grapes look awful, the resulting sweet wine is exquisite, with the multiple sweet flavors lingering a long time on your tongue (i.e., a "long finish").

Trockenbeerenauslese [trawk-uhn-bay-ruhn-OWS-lay-zuh] is a completely different process, but results in a similar sweet wine with a long lingering finish. A German wine label tells you a lot about the wine. The word *Auslese* on the label, for example, means the grape bunches were picked later in the season at their peak of ripeness, thereby guaranteeing a superior wine. The term *Beerenauslese* on the label means that individual grapes were inspected and picked for their ripeness, thus guaranteeing a very superior quality.

Trockenbeerenauslese is Germany 's highest classification for very sweet wines made from specially selected, overripe grapes that are left on the vine to dry nearly to the point of being a raisin. The grapes are usually infected with *Botrytis Cinerea* (the noble rot–*Edelfäule* in German), which shrivels them and thereby concentrates the sugar. At fullest maturity they are very concentrated in flavor and sugar, and produce extremely rich, nectarous wines, frequently with a lot of caramel and honey bouquet, rock fruits notes such as apricot, and the distinctive aroma of the noble rot. The body is viscous, very thick and concentrated, and arguably can be aged almost indefinitely. Although it has a very high residual sugar level, the finest specimens are far from being cloying due to their high acidity. Since these grapes are picked one by one — it takes a person one day to pick enough grapes to make 7-8 bottles—and they are very expensive.

Paula and I have had both of these wines, and they are absolutely incredible, and well worth their very high prices for very special occasions such as anniversaries.

Enough of French Wines 101, and back to my German trip: Our hosts, as well as all of the German people we contacted, were most friendly and cordial. In fact, I had only one frightening experience. I was walking home alone from dinner one night. I stopped at a street intersection to allow a car to pass, when three German youths dashed past me, pounded on the hood of the car, and ran off down the street. I crossed the street, and looked back to see several youths getting out of the car and start running towards me. They may have thought that I was with the others. I knew that I couldn't outrun them, so I just stopped, ignored them, and looked in the store windows. They ran part way up to me, then realized that I was not part of the other group, and returned to their car. As with flight testing, when emergencies occur, it pays to think calmly and clearly, and immediately take the best course of action.

The Bremen Town Musicians

One of the most memorable things about this part of the trip was that I got to see the large, bronze statue of the *Bremen Town Musicians* from the fairy tale—the rooster on the back of the cat on the back of the dog on the back of the donkey.

Our team then returned to the American Embassy in Bonn to prepare our report.

The room in the embassy was very interesting. It was a "secure," sealed room. The room was always locked. There were heavy curtains over the windows (I wondered why there were even windows!). There was a room inside this room. This inner room was made of transparent Plexiglas. The Plexiglas floor was above the wooden floor about a foot, supported on clear, Plexiglas "feet," and the room had a separate ceiling, also of clear Plexiglas. The Plexiglas walls were doubled, with clear Plexiglas baffles inside. Even the door hinges were clear Plexiglas so that any small wires would be readily apparent. There were overhead fluorescent lights in the room, a long wooden table, and wooden chairs with arm rests. A separate air conditioning system was provided, and a "white noise" generator outside the room masked any sounds that might have gotten through the heavy curtains inside the inner room.

Incredibly, the electronics expert on our team announced that he could bug this room in five minutes!

"How?" We demanded.

"I could pull up the arm of one of these chairs, and insert a small microphone and transmitter in the mounting hole," he explained.

"But you couldn't have much power in such a small transmitter," we countered.

"Don't need much. I just need to transmit into the fluorescent lights, and I can park my van under any power line within a mile radius of this building and hear every word you are saying!"

And *that* was 1964 technology.

After we analyzed our data and prepared our report, we presented our findings to the German Ministry of Technology. This was in a large room where we were seated at a long, u-shaped table with our American team on one side, our German counterparts across from us, and the interpreters seated at the end. Each of us presented the findings relative to our individual disciplines. After each five-minute oral presentation, the interpreters would translate it into German. They did a remarkable job!

When I presented my flight test part, I mentioned something about tests for lateral-directional instability, typical of flight at these very slow airspeeds. This motion is a combination of roll and yaw, and is very critical to safe flight control. However, I used the vernacular expression that we used at Edwards, and referred to it as "Dutch Roll."

When the interpreter got to this part, the Germans across the room all raised their eyebrows and demanded: "*Dutch Roll*?? Vas is dis '*Dutch Roll*??"

I had to explain that the term referred to the undulating motions of a Dutch ice skater, and got up and "skated" around the room in front of these dignitaries to explain what I was talking about.

Big laughs.

What fun!

I had a small German phrase book that I referred to frequently. I knew how to order beer please ("bier, bitte") and ask for directions to the men's room ("wo die Toilette der Männer ist, bitte?"). And I even asked for my hotel room key in German. It was a most difficult room number: Room 45. ("fünfundvierzig). The young women behind the hotel desk thought my attempts at asking for my room key were hilarious, and refused to give me my key until I had asked for it in German.

What fun!

I had written orders permitting me to carry the classified **SECRET** VSTOL information back to Edwards. It would be double wrapped in accordance with Department of Defense directives—the inner wrapping would be marked **SECRET** with DoD references, and the outer wrap would be plain, so as not to draw attention to it. The theory was that if someone got the package by accident, and opened it, they would be encouraged to return the inner package unopened. Of course I had to keep the package in my possession at all times, even when going to the bathroom on the plane.

At Edwards, before leaving, our security people asked me if I wanted to carry a pistol. It would be authorized because I was traveling with DoD classified documents. I was not afraid of guns, and knew how to fire a pistol, so I thought carefully about the offer. If I carried a pistol, I would have to be prepared to shoot someone. I didn't want to have to make that decision in the heat of the moment, so I declined.

But what should I do when the US Customs Inspectors went thorough my luggage and briefcase back in New York and saw this sealed package? I asked the US Consulate: “What do I do?”

“They can’t see this information. Show them your orders.”

“That’s not my question! What do I do?”

“They can’t see it!”

“Yes, I know! But what do I *do*?”

Finally he answered my question: “The customs inspectors are not cleared for secret, so you would have to have an FBI agent come and open the package, if it comes to that.”

“Thanks. Now I have a backup plan.”

Sure enough, in New York, when the customs inspector got to my sealed brown paper package, he asked what was in it.

“That’s my DoD classified information. Here are my orders permitting me to carry it. You can’t open the package.”

He looked at me, glanced at my orders, and then at the package, several times. He held the package in his hand, bouncing it up and down a few time, listening to see if it rattled or gave any clue as to what it contained. Finally, he just put it back into my briefcase, put all of my belongings back into my suitcase, and told me to go ahead. I guess he figured that if I was carrying anything illegal it would be in that little 8 ½ x 11 inch package. I have no idea what would have happened if this had occurred after the 9/11 Twin Towers attack.

The flight home was the longest day of my life! I got up at 6:00 AM Frankfurt time, flew to New York, changed planes and arrived in Los Angeles at midnight west coast time, where Paula picked me up and drove me to her folk’s house in Van Nuys. I finally got to sleep about 2:00 AM, Los Angeles time. That day I was traveling for 29 hours.

But what a great trip!

And I learned enough to be able to kill a bad VSTOL development program that was being jointly developed by the U.S. DoD and German Ministry of Defense in 1967, and that had the full support of the U.S. Secretary of Defense Robert S. McNamara!

"Bluefly"

As the Air Force's VSTOL Flight Test Engineering expert, I was assigned to a program that required a special "sensitive" clearance from the Foreign Technology Branch. The program would be to evaluate any VSTOL aircraft that was "obtained" from Russia or Soviet Bloc countries. My job would be to work with the other DoD specialists to evaluate its performance, technology, and combat capabilities. I had a special passport and had to take worldwide immunization shots.

The sensitive aspects, as explained in my special briefing, was that if the Russians even *knew* that we had the aircraft, they would conduct an inventory to find which was missing, and probably shoot those individuals whom they thought were responsible!

The procedure was this: I would receive a mysterious telephone call, with the one, cryptic word "Bluefly." I would immediately stop whatever I was doing, tell my boss and family that I would be gone "somewhere" for an indeterminate time, and fly to Wright-Patterson AFB, Ohio, to rendezvous with the rest of the team.

It would have been great fun, but the call never came.

Later, while studying STOL and VSTOL aircraft suitability at American Airlines, and working closely with the Federal Aviation Administration in Washington, D.C., and NASA Langley, Virginia, I met the Czechoslovakian pilot who defected to the U.S. with a Russian Mig-15. He was on the American Aerobatic Team, but could not compete at the Paris Air Show. He could not go out of the U.S. because the Russians had put a price on his head!

Such was life during the Cold War.

The XC-142A Tilt-Wing VSTOL

In the early and mid 1950s, industry and NASA did considerable analytical, wind tunnel, and flight research on many types of vertical and short takeoff and landing (VSTOL) aircraft concepts. All of the services were interested, because they wanted air vehicles that could takeoff, hover, and land like helicopters, but had the speed and range of conventional airplanes. By the late 1950s, several concepts had proven worthy of further development and military evaluation.

Thus was born the Tri-Service VSTOL program. Three experimental VSTOL transport aircraft were developed for military flight evaluation: the XC-142A tilt-wing, built by Ling-Temco-Vought (LTV), teamed with Hiller Helicopters and Ryan Aeronautical; the X-22 tilt-ducted-fan, tandem wing, built by Bell Aerospace; and the X-19 tilt-prop, tandem wing, built by Curtiss-Wright. The Navy's X-22 was tested exclusively at Bell's Buffalo, New York, facility, and eventually became a very successful variable stability flight test resource at Calspan.

In the early 1960s, a Tri-Service VSTOL Test Force was formed at Edwards for Category II (performance, and stability & control) and Category III (operational suitability) tests of both the XC-142A and X-19. Since these were weird aircraft, naturally they were assigned to me, and for a year or so, I *was* the test force at Edwards, writing test plans, planning instrumentation, organizing the new test force, and planning and attending interminable meetings. My work in this capacity earned me a Civil Service Sustained Outstanding Performance commendation.

My first on-hand experience with the XC-142A was at the cockpit mockup meeting at LTV in Dallas in 1962, where the biggest decision was whether the pilot should sit in the left seat, like a conventional aircraft, or in the right seat, like a helicopter.

Conventional pilots sat in the left seat and the co-pilots in the right seat for several reasons, most obviously because a single set of engine controls could be placed in the center, convenient for both pilot and co-pilot.

Helicopter pilots sat in the right seat and the co-pilots sat in the left seat because neither wanted to ever take his hand off of the cyclic control stick that controlled the dynamically unstable helicopter's attitude. They controlled power and height through a collective control lever by their left sides, which didn't take continuous control movements and could be locked into place during cruise.

In the XC-142A, the helicopter people won out, and located the pilot in the right seat, and the copilot in the left seat, much to the dismay of the fixed-wing pilots.

The XC-142A's Wing Points Straight Up and It Flies Backwards

The XC-142A looked like a railroad boxcar with short, stubby wings. Its mission requirements to carry an Army truck, but fit on a Navy carrier's elevator, dictated its boxy shape. It was powered by four GE T64-GE-1 free turbine turboprop engines, swinging 15-foot, four-bladed propellers. The requirement to have the entire wing covered by the prop wash, to prevent wing stalls during conversion and STOL operations, resulted in the wings not extending past the outboard propellers. To preclude loss of airflow over the wing during STOL, or loss of thrust from a propeller during VTOL and hover in the event of an engine failure, a cross-shaft was provided between all four propellers. This meant that all four propellers began turning when the first engine was started. For longitudinal control in VTOL and hover, a small propeller was installed in a horizontal plane at the tail, just behind the empennage, driven by a shaft through the fuselage, connected to the main propeller cross-shaft at the wing root.

XC-142A during vertical takeoff and conversion to cruise flight

During VTOL and hover, the wing was tilted vertically, and lift-off was made by raising a helicopter-type collective lever on the left side of each pilot. Longitudinal control was commanded to the tail prop by moving the control stick fore and aft. Roll control was commanded to starboard and port propeller thrust differentials by lateral stick movements, and yaw control was commanded to aileron deflections by the rudder pedals. In this configuration, the ailerons were in the high velocity propeller wash, and were very effective.

When your aircraft has a wing that can point straight up and can fly backwards, you have to expect weird things in its flight control system.

After liftoff, a conversion to conventional flight was made by simply pushing the wing-tilt switch forward on the collective lever, the wing came down and the aircraft converted to a real airplane in about ten seconds, flying at 140 knots.

An ingenious, fully mechanical box of cams and levers in the flight control system sorted everything out during transition, and was completely reliable and trouble-free throughout our test program.

Airplane flight used normal aerodynamic controls.

"Chief Engineer? Sounds Like You're Driving a Train!"

Our test force would comprise Air Force, Army, Navy, Marine, and West German participants (West Germany was also developing VSTOL concepts). I had to set up the test force organization to satisfy, not only the many test objectives, but also to satisfy the various services that they were acceptably represented. Therefore, in order to assign Army and Navy "project" engineers, I designated myself as "Chief Flight Test Engineer."

When the Test Force was finally on-site and operational, Colonel Polve, chief of flight test engineering, and my administrative boss, noticed this, and didn't like it. He called me into his office.

"What's this 'Chief Engineer' thing? You sound like you're running a train!"

"Well, I wanted to be able to assign..."

"You should be 'Project Engineer,' not 'Chief Engineer!'"

"Well, the Army and Navy engineers..."

"I'm going to get to the bottom of this! I'm going to call Jesse Jacobs!"

He ignored my attempts to explain, and called Lieutenant Colonel Jesse Jacobs, the Tri-Service Test Force commander. He explained his concern, and listened to Jesse's reply. His expression changed slowly from anger to complete consternation.

"I see," he finally said, and hung up the telephone. He stared at the phone for a moment, then turned to me and exploded:

"You're *'Chief Engineer'* because *you* made up the organization!!"

"Yes sir."

"But.. .but..." he sputtered.

"Is that all, sir?" I asked, respectfully.

When I left, he was still staring at his telephone in disbelief, but he never said another word about it. I didn't even have to explain that the title "Chief Flight Test Engineer," instead of "Project Engineer," got me the GS-13 promotion instead of a GS-12.

A Lot Of Data In Only Ten Flight Test Hours

In 1964, the XC-142A was slowly progressing through its Category I tests at the LTV plant in Dallas, Texas, was way behind schedule, and I was concerned that it would be many months before we had any Government test data. I was also concerned about its suitability for Category II testing because of the many maintenance problems the aircraft had encountered during Category I. I expressed this concern to Major Lou Setter, Chief of Performance Branch, and my immediate administrative boss. He agreed.

"The Navy participates in contractor Category I tests, during its Navy Preliminary Evaluation. How about an MPE—Military Preliminary Evaluation—at LTV?"

This sounded good, ASD liked the idea, and I wrote a ten-hour flight test plan to be performed at LTV in Dallas. Our evaluation would also include monitoring the maintenance requirements to indicate what was in store for us at Edwards. Our recorded takeoff, chased by a T-37 airspeed calibration airplane from Edwards, was followed immediately by a check climb to altitude. At the top of the climb, we stabilized on speed-power points while the chase plane calibrated our airspeed system. I had installed both performance and stability and control instrumentation on both Category II aircraft, so after each speed-power point, we did aileron rolls, pitch pulses, and rudder kicks, then did a check descent to the next test altitude. Finally, the flight was completed with a check descent to the field and a recorded landing.

We discovered that the aircraft was basically easy to fly and that hover performance was 12 % less than predicted. We identified three safety of flight discrepancies, 22 deficiencies that would curtail our Category II tests if not corrected, 45 deficiencies that, while not needed to allow Category II flight tests, should be corrected and evaluated before Category II is completed, and 43 deficiencies that should be corrected for an operational aircraft.

Our 120-page report documented the status of the aircraft, gave ASD preliminary information on the concept, and identified precisely what had to be done to prepare the aircraft for Category II and III tests at Edwards. A great job by everyone involved.

Time and money well spent.

The XC-142A Comes to Edwards

The XC-142As came to Edwards in 1966, along with a large contingent of test personnel. This included our test force commander, Lieutenant Colonel Jesse P. Jacobs, Test Director; Army Major Billie Odneal, Deputy Test Director; Navy Lieutenant Commander Ralph Bennie, Deputy Test Director; and test pilots USAF Major Gay Jones, Army Major Bob Chubboy, and Navy Lieutenant Commander Bill Casey and Lieutenant Roger Rich. An additional USAF pilot, Major Sam Barrett, was provided from the potential using command in order to evaluate how the aircraft might be used in actual service.

I was Chief Flight Test Engineer in charge of three civilian engineers: Jim Satterwhite, Deputy Chief Project Engineer from the Army; Clen Hendrickson, Project Performance Engineer from our own Edwards organization; and Harry Down, Project Stability and Control Engineer from the Navy. Frank Lucero, from Edwards was responsible for the Operational Suitability engineering tests, and Fritz Bröcker was there from the German Ministry of Technology. There were also other engineering aids helping us with the data analysis, Edwards instrumentation personnel, and many maintenance people. It was an incredibly well qualified, cooperative, and effective flight test team.

There was a group of LTV people on site, also, headed by Rick Riccius for a short time and then replaced by Tom Shepard. Our family and Tom's family became great, life-long friends, and we continue our friendship to the present day as I'm writing this.

This test force was a very social group, and we soon started having monthly dinners, hosted by one of the families. This was, without question, the most enjoyable time of our stay at Edwards because of the camaraderie of the people and the social gatherings.

At Edwards there was sometimes a good-natured rivalry between the test pilots and test engineers. Hollywood made movies about the test pilots, the pilots got all the publicity, and we engineers accepted that. There was a mutual respect and friendship between us, but sometimes we had some pretty lively—sometimes even vociferous—arguments. For example, one day the Army test pilot, Bob Chubboy, and I had a very strong interchange in our office trailer. Finally, he stormed out of the room and slammed the trailer door.

Years later during lunch one day, when he worked at the Federal Aviation Administration in Washington, D.C., and I worked across the street at NASA Headquarters, we remembered that argument.

"When you slammed the door as you left, I knew I had won the argument." I told him.

"I *knew* I shouldn't have slammed that door the instant I did it!" He laughed.

People Skills

After I left Edwards a popular book was published entitled *All I Need To Know I Learned In Kindergarten*. When I read this book it reminded me of one day when I got home from work. I asked my five year old son, Key, what he did that day at kindergarten.

"We played games, then I drew some pictures, the teacher read us a story, and we had milk and cookies." He announced, proudly. "What did you do today, Daddy?"

"Well, I met with some flight test people, than I drew up some graphs of flight test data, I read some test reports, and then the snack wagon came around and I had coffee and donuts."

"Wow, Daddy," Key's face lit up with excitement. "That sounds like fun!"

Then it hit me: five-year-old Key in kindergarten, and me, a graduate aeronautical engineer flight testing the Air Forces' latest aircraft, were doing the same thing!

One of the "lessons" from the book was to treat people with dignity and respect. I had learned that years before the book was published, and I used it effectively one time when I had a personnel problem.

"Ray White," I'll call him, worked for Clen Hendrickson, but was not pulling his weight. Ray was a staff sergeant, but had a PhD in mechanical engineering. He dearly wanted to be up at the Rocket Test Facility working on rocket engines, but was "stuck down here with airplanes." Ray was on flight status and got extra pay.

Clen came to me one day, and told me that he needed me to talk with Ray. I set up a meeting with Clen and Ray for the following day, which would give me time to figure out what to do. Fortunately, the Air Force had sent me to several technical short courses and several on sensitivity training and interpersonal relationships. This Management and Executive Development course proved very useful to me through out the rest of my life.

One story told at the three-week Executive Development Course given at Texas A&M stayed with me: Bill Oncken told the story of how he had given a lecture on the importance of keeping your focus on what is really important, not just what is expedient at the time, but on your long range goals.

He said that several months later he ran into a former attendee who was president of an oil company in the Middle East. The oil company president told Bill that the lecture had saved his company. The details are important enough to repeat here:

Communist sympathizers were very active in the Arab country, and the president knew that they wanted to shut down his plant. He said that when he returned from Bill's lecture, that hammers and scythes–Communist emblems–were splashed all over the front of his company headquarters building. His first reaction was to call in the culprits–he knew who they were–and lay into them, probably firing them. But he also knew that this was exactly what they expected, and would give them the justification for demonstrating, holding huge rallies, and perhaps shutting down the plant. So the president bided his time for a day, figuring out what to do. Then he called in the three men.

"I saw what you did to the front of this building," he accosted them. They looked at each other expectantly, waiting for what they thought was coming.

"As you know, I don't like this sort of thing!" They smiled at each other, ready for the onslaught.

"But since these things are important to you and others who work here, they are important to me."

The three men shifted uneasily in their seats. This wasn't what they expected.

"And the paint, dripping down the front of the building doesn't really look very good, does it? It looks sloppy and doesn't give your cause a very good image. Now, does it?"

"Well, no," they agreed, looking at each other in confusion.

"I'll tell you what: If you will draw up some nice looking posters, I'll have the graphics department print them up and you can post them around the building. How about that?"

Wow! This knocked their socks off! They didn't expect this from a man known for his violent temper and his explosive reactions to things he didn't like! But this was certainly a reasonable offer and they could hardly refuse. So they agreed, and in due course they presented their banner design, which, somehow, never really got printed or displayed.

"You saved my company," the president concluded. "Thank you."

I have remembered this story ever since, and it helps me to think things through before acting, and to consider what I really want to accomplish in the long run. To make good decisions instead of bad decisions, and ultimately to make more right decisions than wrong decisions.

All of these lessons-learned helped me plan my approach to Clen's problem with Ray White.

So the next day I sat down with Clen and Ray, and told Ray that Clen had told me that he (Ray) was not doing his job. I told Ray that I understood why he was depressed, but that it was his own fault for not taking ROTC and earning a commission. I told him that I knew he wanted to work on rocket engines, but that hadn't happened. I told him that he had a special privilege by being on flight status, but that it wasn't fair to the others for him to slack on his work. I asked him if he was really satisfied with his performance, and he agreed that he was not. I let him talk for a few minutes, to express his feelings and desires.

Then I told him that I saw two courses of action: either he could straighten up and work harder, and I would leave him on flight status, or he could continue doing unsatisfactory work and I would take him off flight status. I asked him if there was any other course of action that he could think of. He said no.

I told him to decide which course of action he wanted to follow and that I would support his decision. In any case, I would continue trying to get him reassigned to the rocket test site.

The next day, Clen told me that Ray had decided to straighten up, and he did improve some. Eventually we got him reassigned, so he was happy. I asked Clen what Ray's reaction was to our meeting, and Clen told me that Ray was ashamed, and felt that he had let me down.

Yes!!

Treating people with dignity, and respecting their situation and feelings is well worth the effort.

Unfortunately, not all personnel management events turned out so well. Two incidences come to mind. The first involved one of the military maintenance personnel when we took delivery in Dallas of the first XC-142A to be delivered to Edwards.

The person in question, we'll call him "Charley," had been having a fling with a woman during his trips to LTV. She had a second floor apartment, that had once been a motel, and had given him a key to her apartment.

When Charley finished his first day at LTV preparing for transfer of the aircraft, he went to her apartment. It was late and he thought he would surprise her. He quietly let himself into her apartment. She was already in bed, so he sat down on the edge of the bed and tapped her on the shoulder. To his great surprise, a strange young woman sat up and screamed, and her husband raced out of the bathroom!

Charley bolted for the door, and leaped the railing, forgetting that it was on the second floor with a concrete courtyard and swimming pool. Unfortunately, he missed the swimming pool and broke his leg on the concrete. Colonel Jacobs spent half his time at LTV arranging transfer of the aircraft and the other half of his time bailing Charley out of jail, and convincing the newly weds, whom Charley had surprised, that they shouldn't sue! He was successful on both counts, and they never told us who it was who had leaped the railing.

But only one person came back from Dallas on crutches.

The other incident involved me. As the supervisor of the Tri-Service group of engineers, I was responsible for preparing their annual performance reports, and one of the non-Air Force engineers came to me with his evaluation forms. I always tried to do an honest and thorough job, so I would reproduce the forms and fill them out on three consecutive days with out looking at the other entries. After the third review, I compared the entries. I kept those that were consistent, and carefully studied those that had differences. This way I figured I would avoid any quick reactions.

On this particular one, I gave the engineer—we'll call him "Charley," but this was a different Charley—high marks on all except ethics. I gave him a fairly high mark, but not the highest. Here's why:

On trips to LTV, traveling on Government orders, we had to economize and share hotel rooms, two to a room. Charley and I had shared a room on one such trip. We had checked in, but after work I borrowed a car from LTV (they were very generous about that) and drove over to Fort Worth to have dinner with Mother. I returned to the hotel about midnight, opened the door, and found Charley sitting on the bed with a waitress from a local restaurant where we ate breakfast. Charley was married.

All three of us were embarrassed. It was too late for me to find another room, so Charley had to take his paramour home. When I prepared his evaluation, in good conscience I could not give him the highest marks for ethics. I figured that if he would lie to his wife, he might not be completely honest on other matters. I'm not a prude, and never mentioned this to anyone else. That was purely between Charley and his wife, but this time it also involved me. He was upset, but he should have thought of that himself.

Fire Bottle Test

I especially appreciated Colonel Jacob's confidence in me. He was one of the best bosses I ever had. We had two airplanes at Edwards, one was heavily instrumented and we were getting only about one flight a month because of the difficulty of maintaining the aircraft and the complex instrumentation. The other aircraft was not instrumented, and was flying regularly for the operational tests — basically just flying around, landing on various surfaces, and demonstrating simulated missions so we could figure out what the aircraft concept could do and what it couldn't do.

Two engineers from Wright-Patterson AFB came to Edwards one week to conduct a special test. They wanted to place a "fire bottle" in one of the engine nacelles, fire it, and photograph the patterns the fire retardant agent made in order to evaluate its effectiveness in putting out an engine fire.

Colonel Jacobs called me in and told me that he wanted to use the instrumented aircraft because the un-instrumented aircraft was flying some tests off base and would not be back until the end of the week.

"I don't think you should use the instrumented aircraft. I don't know what the agent might do to the instrumentation, and besides, we aren't flying this aircraft enough as it is."

"Rob, these engineers will have to sit around doing nothing all week until the other aircraft gets back. I don't want to have to keep them here all week."

"My advice is 'don't use the instrumented aircraft!'" I don't know how much of this was engineering analysis and judgment and how much was just pure intuition. I couldn't really justify my recommendation on any known facts or experiences, but I felt very strongly about this and didn't really know why. Maybe it was pure intuition. If it ain't broke, don't mess with it!

I left the office, thinking that I had lost that argument, but it turned out that Colonel Jacobs, while not really understanding or agreeing with my recommendation, respected my judgment, even though it was based almost solely on intuition, and told the engineers to wait for the other plane.

They waited until the un-instrumented airplane returned at the end of the week.

They installed the fire bottle in the number two engine nacelle.

They set up their cameras.

They started the cameras and fired the squib on the fire bottle.

The fire bottle exploded and blew a two-foot diameter hole in the aircraft wing!

The next day we flew the instrumented aircraft.

Why Correct Your Data When You Can Test At Sea Level Standard Conditions?

I had installed both performance and stability and control instrumentation on our Category II aircraft in order to get the most test data from each flight. Before we could begin our Category II performance flight tests at Edwards, however, we had to figure out how to reduce the performance data to standard conditions. Should we use airplane data reduction methods, or helicopter methods? OK, helicopter methods for VTOL and hover, and airplane methods for cruise flight—but what about STOL and transition?

Helicopter performance data are obtained using density altitude, rather than pressure altitude like conventional airplanes, and the data are presented in coefficient form of C_P vs. C_T (power coefficient versus thrust coefficient). One chart is all that's needed for all flight conditions. But would this work for tilt-props, and what about the tail prop? Sometimes the tail prop thrust was up, and sometimes it was down, and it would fluctuate continually with pilot longitudinal control movements.

I know! I'll just ignore the tail prop!

This solution wasn't as cavalier as it may sound. What I was really doing was assuming that the C_P/C_T relationship of the tail prop approximated that of the main props, and that any differences would be negligible because the total thrust of the tail prop was a very small percentage of the total thrust of the main props.

I didn't mind interpolating hover performance for conditions between those that we had tested, but I did not want to extrapolate the data into areas that we had not tested. In other words, we needed to attempt to hover at the heaviest weights possible in order to get the most data. If you don't have the power, you can't hover. Everybody knows that.

The GE T64-GE-1 engines were design rated at 2850 SHP each, but every engine was exceeding that power on GE's test stands. Therefore, I requested the power test data for the entire fleet of T64 engines, by serial number, and hand-picked the eight engines for our Category II performance and stability & control aircraft. These engines provided from 3200 to 3400 SHP each. We had two real hotrods!

Helicopters and VTOL aircraft have pronounced ground effects that vary with disk loading (pounds of aircraft weight per square foot of total rotor swept area or thrusting device areas). For helicopters, this is strongly positive—for jet VTOLs this can be strongly negative. We weren't sure what it would be for the XC-142A, but I suspected that it would be very positive.

For these hover in-ground-effect tests, the XC-142A pilots needed to know how high the wheels were off the ground, so I designed a "Hover Meter" and had Instrumentation Division build it. It consisted of two twenty-foot poles, one with two round disks located at the top and two feet below the top, the other had two cross bars similarly located. Placed a measured distance in line in front of the XC-142A, the pilot could lift off to various wheel heights (6 inches to about 10 feet) that were indicated to him by aligning combinations of bars and disks. Sounds crude, but it worked fine.

XC-142A Hovering in front of the "Hover Meter." Note the circles on the near pole and bars on the distant pole

For our hover tests, I sent the aircraft to Naval Air Station Point Mugu, California, in May, because at that time of the year runway conditions would be very close to standard atmospheric conditions–sea level and 59 degrees Fahrenheit in the mornings. I had the aircraft loaded to a weight that was above our maximum hover weight, which required special approval from LTV. We set up our "Hover Meter," and started our tests.

The aircraft struggled into the air, attaining a wheel height of only two feet. After burning out some fuel, we got up to 30 feet, which was essentially out of ground effect. All of this at essentially sea level, standard conditions. This proved that the XC-142A had a very strong positive ground effect in hover.

After STOL takeoff and landing tests, we returned to Edwards to study the data.

Sam Barrett, the Air Force "user" pilot on our Tri-Service Test Force, was most grateful for the hover tests at Point Mugu—if he had not been at Point Mugu, he would have been on the XC-142A in Dallas that crashed during a maximum rate re-conversion to hover test, killing the entire LTV flight test crew.

Predicting In-Ground-Effect Hover to Within Inches

Another argument with my administrative Boss. He wanted to go immediately to Bishop, where Edwards did its hover testing at 5000 feet density altitude, but I held out for reducing the data first, so we would know what we were doing. I won out, and our team had two days' respite while I worked out the C_P/C_T plots, and drew the in-ground-effect hover charts.

For these hover tests we took almost the entire test team to Bishop. We would have to do these tests at Bishop's little general aviation airport early in the morning before the winds picked up. The aircraft was very loud, so loud, in fact, that the pilots had to wear noise-canceling headphones (135 PNdB in the cockpit, where 86 PNdB starts hearing loss, and 120 PNdB is the threshold of pain!) I was concerned about disturbing the local residents and perhaps hurting the rapport that the AFFTC had established with them for further helicopter testing at the little airport. So I sent Dal Taylor to Bishop the week before to talk to the mayor and to put an article in their weekly newspaper about our forthcoming tests.

The message was this: "We need to test this new type of military aircraft at this high altitude test site. This is important to our military security because this new concept could help our armed forces fly into unprepared fields. Unfortunately this is a very noisy aircraft, and we must test early in the morning to get good test data. The noise may wake you up. We are sorry about that, and ask that you put up with us for a few days. If you are interested, we would be happy for you to come out to the airport and watch our tests. It will be very interesting, and we will show you where to stand so that you will be safe and won't be in our way. We appreciate your tolerance."

Several townspeople actually came out to watch these early morning tests, and we never got a complaint. Dal did a good job!

At Bishop, we fueled the aircraft to its maximum weight, and set up the "Hover Meter." The first liftoff was to a maximum power hover at a wheel height of only 6 inches.

I plotted the point on my charts, and asked the pilots to fly a STOL circuit to burn off 1000 pounds of fuel, which they did.

"You should be able to hover at 12 inches this time," I radioed the Army pilot, Billy Odneal.

He lifted off to one foot wheel height.

"Burn off another 1000 pounds of fuel, and you should be able to hover at 18 inches."

"No, we should be able to go right up to five feet or so, this time," Billy replied, as he flew another STOL circuit.

"Eighteen inches," I corrected, with confidence. I was looking at the real time plot!

Billy lifted off. "How high are you, Billy?"

"Well," said Billy, squinting at the Hover Meter, "I'm about halfway between one foot and two feet..."

"How many inches is 'halfway between one and two feet?'" I really rubbed it in!

For the next test, the aircraft lifted off to about seven feet, during which its height stability was marginal. The aircraft would rise up to a little above seven feet, then drop down and the wheels would just lightly kiss the pavement, then it would rise up again. Classical case of maximum hover ceiling at the limits of ground effect.

Based upon our hover tests, I was able to construct a circular sliderule for predicting vertical takeoff performance. The pilot would note the torquemeter reading in the cockpit during preflight run-up of each engine, set this into the sliderule, and, for the existing pressure altitude and temperature conditions (to determine density altitude), and read off the in-ground-effect and out-of-ground-effect aircraft weights at which he could hover or make vertical takeoffs or landings.

I submitted this to the Air Force for patenting, and, on September 15, 1970, received Patent Number 3528605 from the United States Patent Office.

Inventing Data Reduction Methods As We Go

STOL takeoffs took about 3.8 seconds, and could be made with the wing incidence angle anywhere between about 5 degrees (the lowest wing angle where the huge props wouldn't hit the ground), and about 35 degrees. Between 35 and 60 degrees, ground effect recirculation disturbed the aircraft to the extent that the lateral control system could not control it below 25 feet. With all of the variations of power, aircraft weight, wing and flap angles, and atmospheric conditions, how to define the STOL takeoff performance?

After much manipulation of the data, I finally discovered that a specific energy data reduction, $E/(C_P/C_T)$ vs. ground distance, put all of our STOL takeoff data on a single line.

What fun!

Quick test program reorganization

The XC-142A contractor and Government flight test program was rife with crashes—but the #4 aircraft crash in Dallas had the only fatality.

Upon learning of the fatal crash in Dallas, I realized that the Tri-Service System Project Office (SPO) at ASD would want to reassess the entire Category II flight test program, so I immediately began writing a much abbreviated performance and stability & control test plan that I thought would provide us the minimum amount of test data needed to assess the tilt-wing turboprop VSTOL concept for possible military use.

The end of the week we indeed got a request from the SPO to identify the minimum tests needed to complete Category II, and our fully coordinated and approved reply was in the mail within four hours!

Ralph Benny, one of the Navy test pilots, became concerned with the safety of the XC-142A, and suggested that more Contractor Category I type tests be conducted before continuing our military Category II flights tests. Since no one, including our test force director Colonel Jacobs, took any of our concerns lightly, this concern was expressed to the XC-142A System Program Office. After much discussion of all alternatives, and in view of the greatly shortened Category II flight test program, the SPO agreed, and John Conrad, LTV's Chief Test Pilot, arrived at Edwards to conduct a few more Category I type flight tests.

This didn't take long.

On his first VTOL takeoff and immediate, rapid conversion to forward flight, the aircraft started pitching nose up. When John ran out of forward control stick to control this pitch-up, he pushed the collective lever down to reduce power. This controlled the pitch-up, but unfortunately also caused a very hard landing–hard enough, in fact, to cause the wing to crunch down to the zero wing incidence angle, driving the two wing incidence screw-jacks up through the wing and collapsing the landing gear.

In short, it was a mess, but no fire and no injuries.

When no control system malfunctions were found, we began assessing what might have caused the pitch-up. I remembered my helicopter testing experience, and thought about the phenomenon of "translational-lift."

Broken XC-142A—the last XC-142A flight at Edwards

When a single rotor helicopter is hovering, the thrust vector is straight up along the rotor shaft axis, acting much like a propeller. In forward flight, however, the entire rotor disk swept area generates additional lift, much like an airplane wing. This is what allows a helicopter to fly forward at much higher altitudes than it can hover. But the lift from a wing does not act at its center, but at its "quarter chord." The quarter chord is the point, or line, that is about 25% of the distance back from the wing leading edge to the trailing edge. As the helicopter accelerates from a hover to forward flight, and the center of lift shifts forward from the rotor axis to the quarter chord, there is a strong nose up pitch that the pilot counters with a large amount of forward control stick. This is very noticeable when watching helicopters take off.

The XC-142A early design wind tunnel tests did not include the airflow directions associated with this phenomenon in order to save development costs. After this incident, however, LTV conducted additional wind tunnel tests, and, sure enough, the translational lift condition was found to be the culprit.

This was the last flight of our Category II flight test program.

Throughout our test program I sought to determine whether the XC-142A flight characteristics were concept representative or XC-142A hardware representatives. When the XC-142A could not convert from VTOL to conventional mode below 25 feet, due to lateral-directional control problems, at first I thought it a concept limitation. But the Canadair CL-84, a similar tilt-wing, turboprop VSTOL aircraft did not have that problem, I concluded that our problem was an XC-142A deficiency, not a concept limitation.

Canadair CL-84* Dyna-Vert *tilt-wing, turboprop, VSTOL aircraft

What happened to the XC-142A aircraft?

- Aircraft # 1—Fatal crash at LTV during max rate re-conversion
- Aircraft # 2—Wiped out left landing gear during STOL landing at LTV
- Aircraft # 3—Broke in half during hard landing at Edwards
- Aircraft # 2 ½—(Built by combining parts of aircraft # 2 and # 3) Hard landing during max VTO at Edwards, ended Category II program
- Aircraft # 4—Survived to fly at the Paris Air Show in 1966, and then to NASA Langley in 1970. It now enjoys a well deserved rest at the Air Museum at Wright-Patterson AFB, where our Test Force visited it during a reunion in 2002
- Aircraft # 5—Ran into hanger door at Edwards

A Most Difficult Aircraft To Maintain

An advantage of the XC- designation, over the X- designation that it should have had, was that it gave the illusion of being about ready for production. A big disadvantage of the designation was that people expected it to be about ready for production. In fact, maintainability and reliability were specifically not included in the design requirements in order to reduce its development cost. Big mistake. As a result, the mean time before failure was only about ten minutes. Fortunately, most of these failures were minor—bracket cracks, hydraulic leaks, etc.—and were not safety of flight concerns.

Vibration was the major culprit that broke fuel lines, wires and brackets, and cracked structure. LTV set up a special vibration test rig, using aircraft # 4, in which they shook the aircraft at all of the forcing frequencies. With four four-bladed props operating at two different RPM, and a three-bladed prop at the tail operating at another RPM— there was a forcing natural frequency for almost everything in the aircraft!

LTV set up lagged-synched strobe lights, which didn't stop the views of the vibrations, but just slowed them down so they could be seen. With this scheme they could watch the cables and components writhing in slow motion. They would then tag the component with its natural frequency. It was like walking through a forest of snakes, quite a sight!

In the meantime, at Edwards, I was cataloging all of the failures logged by maintenance for our entire Category II and III flight tests. This was before handy little computers, so I had to do it by hand. I made up a box of 3x5 index cards on which I had noted the failures according to system (hydraulic, electrical, flight control, etc.), location (wing, nacelle, tail, center fuselage, etc.), and component (bracket, hydraulic line, wire, structure, etc.).

Gay Jones, the project pilot, and I then went to LTV to discuss what should be done to the remaining aircraft to enable them to safely complete the remaining flight tests.

With the results of LTV's vibration tests, and my box of index cards, we sat down with Gale Swan, LTV's Chief Engineer, and went through XC-142A systems item by item. At issue was which components to “zero-time” and which to just continue to monitor. Zero-timing everything would cost too much money and time.

For example, we considered a wiring harness in the nacelle. The vibration tests showed it in resonance with the four per rev forcing frequency from the prop, my index cards showed a high failure rate for wires in the nacelle next to the props—put in a new one!

How about a control cable in the aft fuselage? It was vibrating considerably, but we had no record of cable failures and the aft fuselage was relatively benign—leave it alone!

The result of this analysis allowed the # 4 airplane, when it went to NASA Langley in 1970, to fly fifteen hours before its first hydraulic leak. An incredible improvement.

XC-142A Primary and Secondary Milestones

The XC-142A flight test program at Edwards set many "firsts." In fact for each primary first, it also set a secondary first at the same time. The following lists these important milestones (with tongue set firmly in cheek): Reading this list provided great amusement at our monthly dinner party.

Date	Primary "First"	Secondary "First"
11 Jan 1965	First XC-142A vertical landing with a military pilot on board	First XC-142A gear-up vertical landing
9 Jul 1965	First XC-142A delivered to Edwards for Category II flight tests	First XC-142A in-flight Category II hydraulic leak
13 Sep 1965	First XC-142A Category II Performance, Stability and Control flight test	Longest period known between flight test and receipt of the test data (about 1 ½ years)
6 Oct 1965	First General Officer to ride in a VSTOL aircraft	First General Officer to refuse to ride in another VSTOL aircraft
4 Jan 1966	First 85% propeller RPM re-conversion and vertical landing	Hardest known landing of an XC-142 aircraft–broke it in half
24 Feb 1966	First large VSTOL transport to operate over an aluminum matting runway	First VSTOL aircraft to be banned from ever returning to a Marine Air Base because it blew away 80 feet of aluminum matting runway
7 Jun 1966	First VSTOL to land on a soft, sod field	First VSTOL to become mired in a soft, sod field
9 Jun 1966	First VSTOL to make a vertical takeoff from the raw desert	First VSTOL to make an IFR vertical takeoff out of a huge cloud of dust
22 Aug 1966	First STOL landing of an XC-142A without the use of the collective power control lever	Longest landing roll in the history of Edwards AFB (ran off the end of the 15,000-foot main runway

Date	Primary "First"	Secondary "First"
3 Oct 1966	Interim Letter Report delivered to the System Program Office on schedule	The only time anything ever happened on schedule in the XC-142A program
7 Oct 1966	Longest non-stop ferry flight of any VSTOL aircraft	First low-level-fuel, emergency landing of a VSTOL at a civilian airfield
25 Oct 1966	First time three XC-142A aircraft were ever in the air at the same time (Field Demonstration)	The only time three XC-142A aircraft were ever in the air at the same time
28 Dec 1966	First XC-142A taxi without using brakes or nose wheel steering	Shortest XC-142A taxi test of the entire program (from the apron into the closed hanger door)
2 May 1967	Heaviest weight hover of a VSTOL aircraft	Lowest recorded wheel height of a pilot induced oscillation
1 Jun 1967	First flight of a US VSTOL aircraft on foreign soil (Paris Air Show)	First gear-up landing of a US VSTOL aircraft on foreign soil

XC-142A Postscript

The maximum allowable payload for a vertical takeoff, dictated by landing gear loads in the event of an engine failure above 25 feet (the lowest height at which a conversion could be initiated), was 4,000 pounds–only half the design payload of 8,000 pounds. I had LTV calculate the added structural weight required to beef up the landing gear to carry the design payload: an additional 250 pounds of landing gear structure would have enabled the plane to carry the full 8,000 pound payload. At Edwards, I couldn't understand why the SPO did not make the design changes that would have made the XC- I 42A concept viable for military use. Then several years later I found out why.

I left Edwards in 1968 to become Development Engineer for VSTOL Technology at American Airlines headquarters in New York City. As part of my studies for airline use of STOL and VSTOL, I met Gene Callaghan, then working for Grumman, but he had been Tri-Service VSTOL SPO chief for a short time during the flight test program. He told me:

"I was ordered to kill the XC-142A and X-19 programs."

"I wondered why you did some of those stupid things."

"Tell me about it! My boss at ASD, Colonel Kirsch, told me to kill both programs. When I told him that I couldn't do that, he transferred me and put in a civilian SPO chief."

It's enough to make you wonder why we do this?

Because somebody's got to.

X-22 Ducted Fan VSTOL Aircraft

Bell X-22 Ducted Fan VSTOL Aircraft in hover and in cruise

Originally one of the three Tri-Service VSTOL aircraft, the Air Force and the Army lost interest (or ran out of money) and decided not to pursue the X-22 concept. The Navy liked the ducted propellers because they are safer than bare spinning propellers on the crowded deck of an aircraft carrier where there are lots of people working near the aircraft. The Navy asked to get its invested money back, but the Department of Defense refused. The Navy then declared that it would pursue the project on its own.

The X-22 flight test program was then managed entirely by the Navy, and flown initially at Bell Aerospace's Buffalo, New York, plant, where the aircraft was built. Later it was fitted with a variable stability and control system that could simulate various VSTOL aircraft handling characteristics, and was used very successfully at Calspan, also in Buffalo, to help develop VSTOL handling qualities criteria.

Although it never came to Edwards, it was nonetheless part of the original Tri-Service VSTOL Program, so its photographs are appropriate here.

X-19–A Pretty Airplane With Ugly Habits

While we were struggling through the XC-142A flight tests at Edwards, the X-19 development was progressing and flight testing was started at Curtiss-Wright across the country in Caldwell, New Jersey.

In 1959, Hank Borst developed a unique propeller VSTOL concept for Curtiss-Wright, that resulted in successful flight tests of the Model X-100. This aircraft was a small, open cockpit demonstrator, with a short stubby wing and a turboprop exhaust exiting the tail to which diverter vanes had been installed for pitch control in hover. The propellers were most unusual.

Curtiss-Wright X-100 VSTOL Demonstrator. You can see the fat propeller roots in the picture

Normally propellers are narrow at the root, and may be wider at the tips. This is to minimize the "radial force" component of thrust that exists when any propeller's shaft is operating at a positive angle of attack to the air stream. This force acts in a vertical direction, in the propeller's plane of rotation. In other words, it provides lift, but it's also vectored aft, causing drag.

Hank reasoned that if a propeller was designed to optimize this radial force (i.e., the propeller shape designed to maximize the Lift/drag ratio of the radial lift component), a substantial lift vector could be realized, and the wing for a VSTOL aircraft could be made much smaller for improved cruise efficiency. The X-l00 flight tests substantiated this concept, and Curtiss-Wright had Hank design a small VSTOL aircraft, the size and appearance of an Aero

Commander, which was a popular twin engine executive transport of the day. They called it the Model 200.

With the advent of the Tri-Service VSTOL program, in 1962 Curtiss-Wright offered the Government an essentially off-the-shelf vehicle for Government tests. The Government was interested, and paid Curtiss-Wright to install, among other things, ejection seats, two Government-furnished Lycoming T-55-L-5 turbo shaft engines to replace the experimental Wankel Rotary engines of the original design, and to rename the aircraft the X-19.

Curtiss-Wright/Tri-Service X-19 VSTOL Aircraft ready for vertical takeoff

The ejection seats, located side-by-side in the small cockpit, are crucial to this story, as you will see. They were the North American Aviation, LW-2B, which were capable of safe crew ejection from 0 to 500 knots, and from 0 to 50,000 feet height above the ground. The design features that provided this remarkable capability were that the pilot stayed in the seat, to which the parachute was attached, and the parachute was inflated by a ballistic charge immediately upon ejection.

The X-19 was the prettiest little VSTOL aircraft of that era. The fuselage was the size and shape of an Aero Commander, it had tandem wings with four propellers, one at each wingtip, and twin engines in the aft fuselage. Cross-shafts from the two engines drove the four propellers, which could be tilted vertically for VTOL and hover, and rotated down for cruise flight. In hover, differential pitch between the forward and rear props provided longitudinal control, differential pitch between the port and starboard props provided lateral control, and differential pitch diagonally provided yaw control—sort of.

Curtiss-Wright was to fly the initial Category I hover tests at its plant in Caldwell, New Jersey, then ship the X-19 to Edwards for its Category I STOL, conversion, and conventional tests. The Tri-Service VSTOL Test Force would then take over and fly the Category II and III tests at Edwards.

The first flight, a hover test at Caldwell, consisted of getting very light on the landing gear and rolling over, collapsing the right main gear. Subsequent attempts demonstrated terrible control harmony about the lateral and pitch axes, and the pilot's inability to handle the aircraft.

No reflection on the pilot, who had been flying Curtiss-Wright's old B-17 with an extra engine in the nose for testing propellers—but the X-19 was a real handful. The X-100 test pilot wanted more money to fly the X-19 than Curtiss-Wright was willing to pay, so they finally hired a helicopter test pilot as chief test pilot. Realizing that the aircraft also had a non-helicopter flight mode, they hired a fixed-wing test pilot, who took a leave-of-absence from the FAA's NAFEC flight test facility in Atlantic City, New Jersey.

In the mean time, back at Edwards, we were busily planning how we would conduct the Category II and III tests when the X-19 arrived.

Then, trouble! At a flight program planning meeting at the SPO, Curtiss-Wright proposed to conduct its Category I STOL and conventional tests at Caldwell, and said they could save the SPO over $100,000 (big money in those days!) by not having to maintain the aircraft at Edwards.

The SPO loved this idea, but I objected—strongly.

"There is no place to make an emergency landing on the East coast. With the dry lake bed at Edwards, we could land the aircraft immediately anywhere in case of an emergency."

"The X-19 is not really a VSTOL aircraft," Curtiss-Wright countered, "it is a VTOL aircraft. In an emergency, it will reconvert to hover mode and land vertically. It can land vertically in a vacant lot."

"I cannot imagine your completing Category I without a hydraulic failure or a gearbox problem," I argued. "In that case, I don't think the pilots will want to enter a more dangerous flight mode. I think they will elect to eject. That will be the end of the airplane, and the end of the X-19 program."

This time, I was overruled. The X-19 stayed on the East coast for Category I, but at least it was moved from Caldwell to NAFEC. In retrospect, I realized later that ending the X-19 program was just what Colonel Kirsch, at Wright-Patterson AFB, wanted.

The aircraft was designed to fly on only one engine—the two engines were only to provide engine out safety. In fact, to save weight, the transmissions and gearboxes were not designed to take the full power of both engines, since such power was not needed and was prohibited by the flight manual. The flight manual also advised what to do in case of a total engine failure.

"Put the aircraft into a dive, arrest your sink rate over the end of the runway, and touch down at 160 knots."

The flight manual then added, almost as an afterthought:

"...but since the landing gear structural limit is only 65 knots, you're not going to make it, anyway, so just bailout if the engines quit."

So much for emergency landings.

I returned to Edwards, and told my flight test engineers to forget planning for Category II at Edwards. Get ready to monitor the Category I tests at NAFEC. I felt so strongly about this, that I entered into my journal the conviction that the X-19 was going to experience a hydraulic or gearbox failure at NAFEC, the pilots would bailout, the aircraft would crash, and the program would be canceled.

My civilian boss—not a member of the Tri-Service Test Force—disagreed with my assessment.

"Keep your engineers here, getting ready for Category II," he ordered.

"The X-19 will never get to Edwards," I argued. "1f we are to get any information at all, we have to get it during Curtiss-Wright's Category I tests at NAFEC."

I lost this argument, also, but only temporarily. He shortly left on an extended temporary duty to monitor the German VSTOL programs, so no sooner was he on his airplane, than I had my engineers on their airplane—headed for NAFEC—instructed to get the most out of the Category I tests as possible. I believed strongly in the adage that: "It is easier to get forgiveness than permission."

Good move!

Although Bob Baldwin, our project test pilot, made only one or two hover flights, the entire flight test program of 50 flights comprised less than four hours (think about that, for a moment), and these were flown mostly by the contractor crew. Bob did identify the attitude control problem, however. He reported that longitudinal control required very large control stick movements, but roll control was very sensitive, and required very small movements. The problem, then, was control harmony.

The contractor pilot and copilot flew the first conversion flight at NAFEC on August 25, 1965. The X-19 was to make a STOL takeoff from the NAFEC main runway, climb to 500 feet, circle the field at airspeeds of 40 to 150 knots, and return to land on the main runway. Bob Baldwin, Don Wray (the Army test pilot), and Dick Homuth (the Navy test pilot) would chase in an H-34 helicopter, and the NAFEC ground cameras would track the entire flight.

Takeoff and climb out were normal. On the downwind leg, however, the copilot reported a high left rear nacelle gearbox temperature. The pilot apparently thought he was reporting hydraulic pressure, and said "OK."

At the turn to base leg, the copilot repeated his warning, and the pilot, this time, understood that they had a serious gearbox problem, indicated by the high temperature. He continued his turn, but, instead of lining up with the runway for an immediate STOL landing, he headed for a small clearing in the trees, which was closer, for a vertical landing. He was the helicopter pilot, remember, and had virtually no fixed wing experience.

The X-19 disappeared behind the trees, and the NAFEC camera operator, in the tower, asked Bob Reschak, the flight test project engineer, whether he should stop filming.

Bob ordered "Keep the cameras going."

The copilot didn't know about landing helicopters, but he knew that any kind of landing in the trees at 105 knots was a bad idea and could ruin your whole day, so he shoved both throttles full forward. Immediately, the X-19 came zooming up from behind the trees.

Remember that the transmission system was not designed for full power from both engines?

At 390 feet above the ground, the left rear nacelle and prop broke loose, and the X-19 instantly pitched up and rolled left. Now the two engines were driving only three props, which also broke and left the aircraft. Now two engines were running full power with nothing to run, so they over-sped and exploded. It had been less than three seconds since the first prop had broken off, the X-19 was upside down only 390 feet above the ground and rolling at 180 degrees per second, and both pilot and copilot fired their rocket seats and ejected towards the ground.

One half second after ejection, the canopy thrusters fired, and 1.4 seconds later the chutes were fully deployed—still 230 feet above the ground. Both crew landed safely, with probably the only ejection system of the time that could have saved them.

The X-19 continued its rolling plunge and exploded upon impact.

Curtiss-Wright did not know that rolls of oscillograph paper are hard to burn—they char on the outside, but frequently the inside can still be developed. Fortunately, Bob Reschak was on hand to advise them, and the data were subsequently recovered.

If I had won my argument for Curtiss-Wright to conduct Category I flight tests at Edwards, the plane could have landed safely on the dry lake bed, the gearbox problems could have been corrected, and the flight test program could have continued.

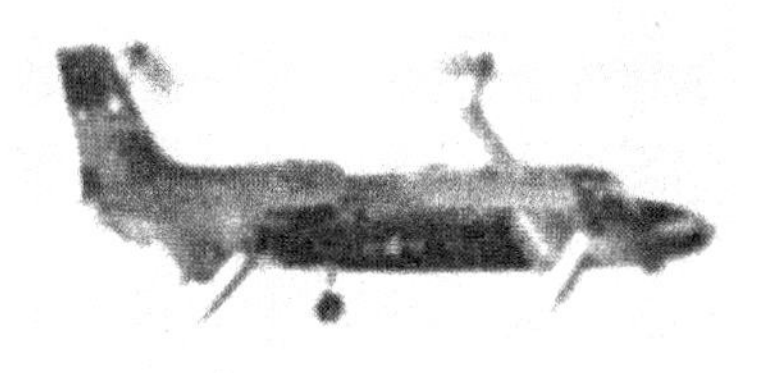

0.0 Seconds ***– Left rear propeller breaks off***

1.0 Second ***– X-19 pitches up and rolls lef***

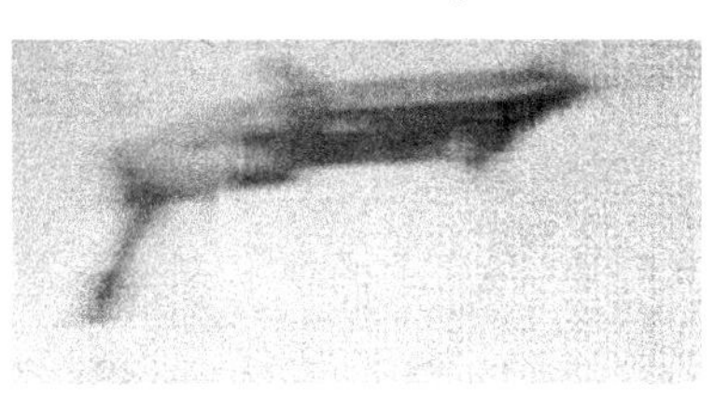

2.0 Seconds ***– All four propellers break off***

2.6 Seconds ***– Both pilots eject through the canopy***

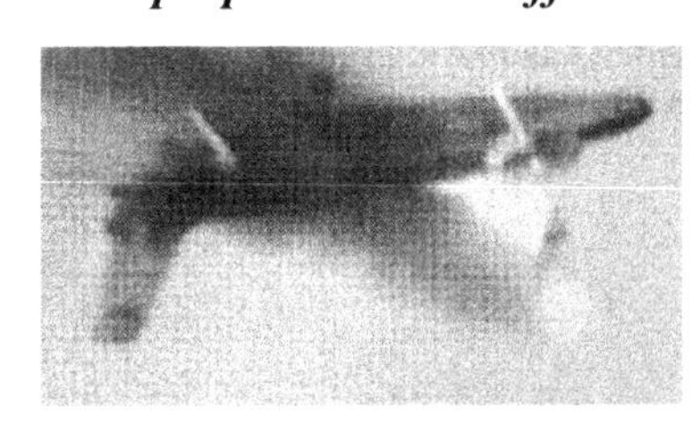

2.8 Seconds ***– Rocket ejection seats clear the aircraft***

3.4 Seconds ***– Both pilots' chutes open***

But the X- 19 never came to Edwards.

The SPO canceled the X-19 program.

The Air Force would have gotten nothing for its investment if I had not disobeyed my boss and sent my engineers to monitor the Category I flight tests.

My civilian boss never mentioned that I had disobeyed him.

I marked the price on my copy of our X- 19 Category I Limited Flight Evaluation—$ 60,000 per copy.

VSTOL Terminology Definition

Bob Baldwin and I were concerned about how to describe the characteristics of VTOL, STOL, and VSTOL aircraft. We needed a consistent glossary of terms so that everyone was communicating effectively.

For example, in this early phase of VSTOL development, a popular demonstration was a vertical takeoff, conversion to forward flight, a circuit of the field, a reconversion to vertical mode, and a vertical landing. This was called a "verticircuit." Note, also, the terms "conversion" and "reconversion."

Clearly we needed specific definitions of these terms. So, to this end, I initiated and formed an Ad Hoc committee to define VSTOL terminology. I co-chaired his committee with Bob, but when he was ordered to Vietnam, another helicopter test pilot, Joe Basquez replaced him and co-chaired the committee with me.

I enlisted our Tri-service VSTOL Test Force to our committee, which included Air Force, Army, Navy, and Marine VSTOL pilots and engineers, and their home organizations. We prepared a first draft, and coordinated it with 24 other NASA, Government, and industry agencies, until we got consensus.

We published *The Report of the Ad Hoc Committee on VSTOL Terminology* in July, 1967, and made widespread distribution.

To Kill A US/FRG VSTOL Program

In the early 1960s, the Federal Republic of Germany (FRG) visited the US aircraft industry, and in 1964 invited the DoD to Germany to evaluate its VSTOL aircraft development programs. At the time, I was the Air Force's VSTOL flight test engineering expert, so, in the spring of 1964, I was assigned to the 13-person team and went to Germany for three weeks.

It was great fun, and we toured the entire West German aircraft industry as guests of the Germany Ministry of Technology. This knowledge, and my flight test experience with American conventional and VSTOL aircraft, and helicopters, came in handy in 1966, when the US and the FRG signed a Memorandum of Agreement to jointly develop a VSTOL aircraft.

Two engineers (I don't remember their names) from ASD visited me at Edwards, to gather information so they could write the performance and handling specifications for the joint aircraft development.

We spent a most interesting afternoon.

"The joint aircraft will probably be a tilt jet concept," they said.

"I agree, because that is what the Germans are doing with their VJ-101 and their DO-31. Basically they are good concepts." I agreed, "but the vehicles themselves have some problems. For example, they've both got way too many engines—VJ-101 has six engines and DO-31 has ten engines. Rolls-Royce loves them, but they are too expensive to build and operate. Both planes are carrying around lots of extra weight that is only used for takeoff and landing."

"The pilots find them easy to fly."

"Did you know that George Bright is the only one to ever fly the VJ-101, and Drury Wood is the only one to ever fly the DO-31?" I asked.

"Really?" They were surprised. "But we saw lots of film of the VJ-101 flying."

"You saw all of the flying that was done," I explained. "The aircraft has only nine hours total flight time, and most of that is at air shows and demonstrations for German dignitaries."

"We didn't know that." They pressed on. "The US/FRG airplane will have an attitude stabilization system."

"Let me tell you about attitude stabilization systems." And I explained:

George Bright, the American test pilot hired by EWR in Germany to test the VJ-101 jet VSTOL aircraft, wanted any pilot to be able to fly it. Therefore, he had the aircraft equipped with an attitude stabilization system—so called "stick steering." If the pilot holds the control stick exactly vertical, the aircraft will be perfectly level. Five-degree left stick, and the aircraft will bank left exactly five degrees and move to the left until it is leveled out.

Piece of cake to fly—except...

This type of stabilization system cannot be engaged on the ground! If the aircraft is not *exactly* at the angle commanded by the control stick, the stabilization system will command engine thrust to attain that angle, up to 100% power from both right wing tip engines, if necessary, in the case of the VJ-101. The aircraft can literally tear itself up on the ground. Therefore, a so-called "squat switch" on the landing gear turns the stabilization system on automatically when the aircraft lifts off.

In addition to this, the VJ-101 was the *only* VSTOL ever built in the free world, that could not be flown, VFR, with all SAS off. It was absolutely uncontrollable. This was not a concept limit—SAS off VTOL was just not a design requirement.

"Picture this," I continued. "These aircraft are planned to be parked around the countryside, on little VTOL pads. They must be ready to takeoff after sitting there for a week or so. The pilot cannot check out his SAS before takeoff, and his aircraft is un-flyable if the SAS doesn't work properly immediately as he lifts off!

ASD was taking copious notes.

"But the VJ-101 has a triplicated, democratic system. If one of the three channels fails, the other two vote it off the line and take over."

I told ASD about my conversion with the EWR people about that when I was in Germany. They were adamant: "It is not designed to fail—it is designed to verk. It vill verk!"

I did, finally, find a German engineer who had at least heard of mean time to failure. But he had the solution: "If the MTBF is 100 hours, we will change the part at 90 hours. No problem!"

I don't think so!

Ironically, shortly after my trip to Germany in 1964, one of the three gyros was accidentally installed upside down. George was making a STOL takeoff, and the instant he lifted off and the SAS came on, it said "Hey, we're upside down!" and commanded full roll control. George grabbed the seat handles to eject, but looked "up" and saw only runway—he counted to "one-half," as he tells it, and then punched out. The Martin-Baker seat got him out OK, but the aircraft crashed on the runway.

"What should we do?" asked ASD.

"Well, I think the attitude SAS is a good idea—that's what the Germans did, so using that is a given for this joint program—but don't you think that the aircraft should be able to take off or land vertically, VFR, with all SAS off? If the SAS fails during an IFR vertical landing, shouldn't the pilot be able to abort the landing instead of ejecting, and fly conventionally to a VFR alternate?"

"That makes sense," they agreed, writing this down. "How about the jet lift, tripod configuration concept? With six engines, if the pilot loses an engine on takeoff, it's only one sixth of his thrust."

"Let me tell you something about this thrust concept."

The VJ-101 had two jet engines on each wingtip, that could swivel from horizontal for conventional flight, to vertical for VTOL and hover. A third pair of fixed engines was mounted in the fuselage, just behind the pilot. The three pairs of engines formed a tripod around the aircraft center of gravity. Variations in thrust provided roll and pitch control, and small tilt variations by the wingtip engines provided yaw control. So far, so good.

Any single engine failure, say a wingtip engine for example, would result in a rolling moment that had to be countered by a corresponding power reduction on the alternate engine. Now this caused a pitching moment that had to be countered by a power reduction in the fuselage mounted engines. A single power failure therefore can result in losing *half* the thrust, not just a sixth!

That's not all.

The hot gas recirculation patterns of the engines in the vertical position precluded high power, pre-flight engine checks. The engines flamed out after only 20 seconds because the hot jet exhaust re-circulated and was re-ingested into the engines. The wingtip engines could not be run to full power in the horizontal position, either one at a time because of the yawing moment, or collectively because of the high forward thrust.

The result of all this was that the aircraft, which had been sitting on the ramp for a week or so, could not be given a good engine run before takeoff, and the engines had to work right the instant they were given full power, or the aircraft would be lost.

"What can we do about this?" asked ASD.

"Shouldn't the pilot be able to either complete a VFR vertical landing or safely abort an IFR vertical landing if an engine fails? Shouldn't he be able to make full power engine checks before takeoff from remote sites? Shouldn't he be able to take off with one engine out, and fly to a maintenance base for an engine change, instead of having to change an engine out in the woods on a parking ramp?"

"Makes sense to us." ASD made more notes.

The result of our discussion was that we completely supported the US/FRG joint VSTOL development agreement signed by Secretary of Defense Robert S. McNamara and the German Ministry of Technology, and ASD specified performance and handling quality requirements that no one could possibly disagree with—but the concept could not meet these specifications!

Months later, Carl Simmons, our AFFTC flight test engineer monitoring the US/FRG program returned from Germany. He was aghast.

"What in the world were you trying to do—kill the program?"

"Yes, as a matter of fact. It's a bad program!" Then I explained the problems, as I had explained them to ASD.

Carl finally agreed that I was right.

"Herr Madelung, at EWR, agreed to those specifications," Carl said. "I told him that his airplane could not possibly meet those requirements."

Herr Madelung was not concerned.

"ASD cannot be serious about those specifications," he said.

The US/FRG program died peacefully in its sleep.

A Real War Story

Our Tri-Service Test Force commander, Col. Jessie Jacobs brought back this story from a visit to Wright-Patterson AFB, Ohio. He had witnessed this in the Officers' Club:

F-104G* Starfighter *for the Luftwaffe

In the mid-1960's, the Federal Republic of Germany had bought the U.S.-made Lockheed F-104G Starfighter to equip the Luftwaffe. A number of senior German pilots were at Wright-Patterson Air Force Base for transition flight training in the aircraft before the fighters were delivered to Germany. The American USAF instructors were also senior officers. Consequently, most of them had flown during World War II—many of them over Europe.

After a day's flying, the pilots frequently gathered at the Officer's Club for cocktails and airplane stories. With twenty years intervening, the war stories were told and appreciated with interest and dispassion—understandable, perhaps, only to those who share the camaraderie of the air.

An American pilot told the story of his dogfight with a German Luftwaffe ME-109 over France. He described the duel in great detail the date, the time, the place, the screaming dives, the roaring climbs, the rolls and tight turns, the blazing machine guns, and the terrible forces on the airplanes and pilots as they both fought for their lives. He described how, finally, the ME-109 burst into flames and the German pilot bailed out.

The USAF pilot concluded his story by saying: "As I brought my plane around to observe, I suddenly realized that the German pilot, swinging helplessly under his parachute, was exactly centered in the cross-hairs of my machine gun sight. All I had to do was press the machine gun button with my thumb. But in that instant, I realized that I had not come to that. There's a difference between killing in war when you have to and murder—and I'm not a murderer. So I rocked my wings in salute and flew back to my base."

The group was silent in appreciation of this powerful story of human compassion, Proof, that even in war, strong men can respect human life and refrain from needless killing.

That they know the difference between war and murder.

Rudyard Kipling put it best in the first (and the identical last) verses of his *Ballad of the East and the West*:

> *"O East is East and West is West, and never the twain shall meet,*
>
> *'Til earth and sky stand presently before God's judgment seat.*
>
> *But there is neither East nor West, border nor breed nor birth,*
>
> *When two strong men stand face to face–*
>
> *'though they come from the ends of the earth."*

The silence was finally broken by one of the German pilots, who said quietly, "that was me."

Last Flight of the XB-70 Valkyrie

The North American Aviation XB-70 Valkyrie was the prototype version of the proposed B-70 nuclear-armed deep-penetration strategic bomber for the United States Air Force's (USAF) Strategic Air Command. Designed by North American Aviation in the late 1950s, the Valkyrie was a large six-engine aircraft able to fly Mach 3+ at an altitude of 70,000 feet, which would have allowed it to avoid interceptors, the only effective anti-bomber weapon at the time.

North American added a set of drooping wing tip panels that were lowered at high speed. This helped trap the shock wave under the wing between the down-turned wing tips, and also added more vertical surface to the aircraft to improve directional stability at high Mach numbers. North American's solution had an additional advantage, as it decreased the surface area of the rear of the wing when the wing tips were moved into their high speed position. This helped offset the rearward shift of the center of lift, or "average lift point," at supersonic airspeeds. Under normal conditions this shift in center of lift caused an increasing nose-down trim, which had to be offset by moving the control surfaces, increasing drag. (In the B-58 we transferred fuel to move the aircraft center of gravity aft) When the wing tips were drooped the surface area at the rear of the wings was lowered, moving the lift forward and counteracting this effect, reducing the need for control inputs. This also allowed the XB-70 to ride its own shock wave.

The XB-70 was huge, in order to carry the enormous fuel load needed for acceleration and cruise at M 3. It had a length of 196 feet, a height at the tail of 31 feet, and an estimated maximum gross weight of 521,000 pounds. It had a crew of four: pilot, copilot, bombardier, and defensive systems operator. The delta wing had a span of 105 feet with six turbojet engines side by side in a large pod underneath the fuselage. The wing was swept at 65.5 degrees, and the wing tips were folded down hydraulically to improve stability at the aircraft's supersonic speeds of up to Mach 3. A large canard fore plane near the front of the fuselage with a span of 28 feet, 10 inches was used for longitudinal stability. In addition to its sharply swept delta wings, the XB-70s had two large vertical tails.

The aircraft was fabricated using titanium and brazed stainless steel honeycomb materials to withstand the heating during the sustained high Mach number portions of the flights. The propulsion system consisted of six General Electric turbojet engines (J93-GE 3) with two large rectangular inlet ducts providing two-dimensional airflow

Although the airplane was huge, the cockpit was surprisingly tiny: not even big enough to stand up in. I know, because I got a grand tour by the North American representative.

General Electric had the publicity idea for a great "family portrait" of the airplanes flying at that time with GE engines: The XB-70 had six, the McDonnell F-4 had two, the Northrop F-5 had two, and the Lockheed F-104 had one.

This was approved, and the fateful day arrived.

GE engines family portrait

On June 8, 1966, we were having an XC-142A flight test planning meeting, when Jesse's secretary dashed in, out of breath, and blurted out: "TheB-70hasjustcrashedandtheyneed C-130piolotstoflyoutothe crashsite!!!!"

All had been going well, until NASA rocket plane test pilot Joe Walker flew too close to the XB-70's wing tip. Because of the sharply swept back XB-70 wing tips and the F-104 wings behind the pilot's line of vision, the F-104 got too close, and was apparently caught in the XB-70's wing tip vortex (von Kármán vortex street).

XB-70 Showing 6 engines

XB-70 with wing tips up

XB-70 with wing tips down

F-104 in flames after midair

XB-70 without vertical tails

XB-70 pitch up

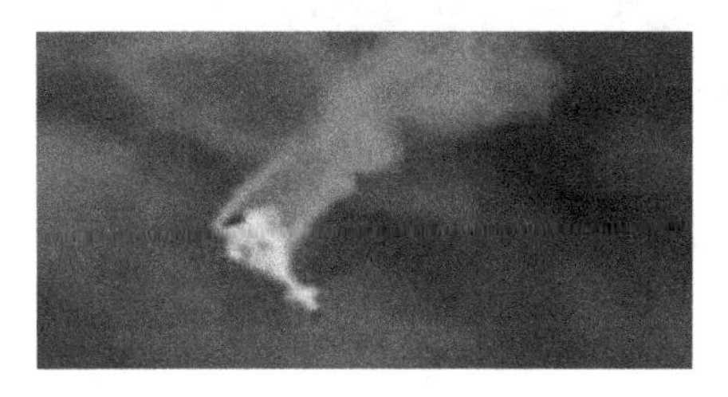

XB-70 in flat spin

XB-70 crash site

In fluid dynamics, a Kármán vortex street (or a von Kármán vortex sheet) is a repeating pattern of swirling vortices caused by the unsteady separation of flow of a fluid around blunt bodies. It is named after the engineer and fluid dynamicist Theodore von Kármán, and is responsible for such phenomena as the "singing" of suspended telephone or power lines, and the vibration of a car antenna at certain speeds.

Since GE had a photo chase plane taking pictures, it is probably one of the most completely photographed crash sequences available.

Joe Walker's F-104 flipped over the XB-70's starboard wing tip, swept across the wing and wiped out both XB-70 vertical tails, breaking both off.

The F-104 exploded into a ball of flame.

The XB-70 flew a short distance, then without the lateral-directional stability of the two vertical tails, went into a flat spin and crashed. The pilot ejected safely, but the co-pilot's ejection seat/capsule malfunctioned, and he was killed with the airplane.

Edwards Farewell

The XC-142A program was coming to a close, and the X-19 program had been cancelled. I had a job offer from the Army Test Facility at Edwards for a GS-15 position, but Paula and I were tired of the Mojave Desert, and flight testing was getting too organized and structured. Even though there were some exciting planes—the SR-71 Black Bird and the US supersonic transport (SST)—but too many people were now involved in every decision. I wondered whether I could have unilaterally done some of the things that I did. For example, I doubt that I would have been able to ignore the scheduled X-19 Category II flight tests at Edwards, based solely upon intuition (OK, and maybe a bit of an educated guess!). A committee would probably have ruled against me and the Government would have gotten nothing from its investment in the program.

Also, Paula and I were tired of the much too casual environment of the desert–there was no place closer than Los Angeles where we could have a formal anniversary dinner. I knew that I could get a job at LTV, based upon my work with the XC-142A, or General Dynamics, based upon my work with the B-58. With my record at Edwards I could probably have named my position at virtually any US aircraft company. All of the other flight test engineers leaving Edwards had done so with no difficulties—basically, decide the part of the country that interested you, contact an aerospace company in that area, and go to work.

I would have mixed emotions about leaving Edwards: many major events had happened to me there, both professionally and personally. I was very proud of my contributions, even more so because Edwards, according to Tom Wolfe in his book about the space program *The Right Stuff*, was "the pinnacle of all of the flight test centers in the world." If I may be permitted to brag a little:

- At my very first performance review as a 6-months second lieutenant, my boss rated me "about the same as the average lieutenant colonel."
- I subsequently received three Outstanding Officer Effectiveness Reports.
- I was offered a regular Air Force Commission.

- I made Captain in five years.
- I was directly or partly responsible for flight test planning that saved four airplanes and possibly their crews:
    - I put water in the B-58 fuel tanks for the refused takeoff tests that kept the airplane from burning when the high-level fuel shutoff valve failed;
    - I suggested asking the aerial refueling tanker to wash the hydraulic fluid off the windshield of our B-58 at 25,000 feet over Edwards so we could land safely;
    - I conducted the hazardous takeoff tests from the sand before the landing that deteriorated the engines; and
    - I recommended that we land the H-41 helicopter immediately after losing a part off the rotor.
- I implemented advanced Tri-Service VSTOL Test Force program planning methods, encompassing flight test, instrumentation, maintenance, and logistics that were followed for many years afterwards.
- I invented a simple device to help the XC-142A pilots judge their wheel heights during in-ground-effect hover tests.
- My special hover tests of the XC-142A at Point Magu resulted in one of our pilots not being on the XC-142A in Dallas that crashed, killing its entire flight test crew.
- I developed new methods of flight testing and data reduction of tilt-wing VSTOL aircraft.
- I developed a simple hand calculator for predicting tilt-wing turboprop VTOL performance that was subsequently patented for me by the Air Force.
- I foresaw the problems with the X-19 program and took action that enabled Edwards to obtain flight test information that otherwise would have been lost when the aircraft crashed during contractor tests and the program was canceled.

- I represented the U.S. Department of Defense in helping to assess the Federal Republic of Germany's ability to develop VSTOL aircraft.
- I initiated, formed, and co-chaired a 12-person ad hoc team of Air Force, Army, Navy, and Marine VSTOL pilots and engineers to define VSTOL terms, and coordinated this through 24 government and industry agencies.
- I had been able to advise the Wright-Patterson Air Force Base Flight Dynamics Laboratory in writing performance and handling qualities specifications for the US/FRG Joint VSTOL Program that helped to prevent a very dangerous VSTOL aircraft from being developed.
- I received a Civil Service Sustained Superior Performance commendation for my VSTOL work.
- I was promoted to the level of Associate Fellow in the august American Institute of Aeronautics and Astronautics for my work with STOL and VSTOL.
- I started the Lancaster Photographic Association, a camera club, that celebrated its 40th anniversary in 2006.

When I was making my rounds, saying goodbye just before leaving, the Assistant Director of Flight Test told me: "You've left some pretty big shoes here for someone to fill." At my going away party, Ray White, whom I had had to discipline during the XC-142A program, bought me a drink and wished me well.

I was very proud of my accomplishment, but I'm most proud of my marriage to Paula McBride, and of our two wonderful children Key and Cheryl. My eleven years at Edwards had been incredibly rewarding, both professionally and personally.

Paula and me

Cheryl and Key

But the bottom line was that flight testing at Edwards wasn't fun any more, and, sadly it was time to move on. So we left Edwards to face our own "Ad Inexplorata." Towards our own unknowns.

New Hire At American Airlines

Paula regularly perused the *Los Angeles Times* newspaper for job opportunities away from the desert, and one day she saw an ad in which a "Leading U.S. airline" was looking for someone with VSTOL experience to investigate VSTOL's possible application in relieving airways congestion in the Boston–New York–Washington, DC, Northeast Corridor.

I sent in a lengthy resume. David Blundell, an American Airlines Lead Engineer, was attending a technical symposium in Los Angeles, and scheduled an interview. The interview with David went very well, and in the summer of 1967 I received an invitation to fly to New York for an interview with Frank Kolk, Vice President of Development Engineering, at American Airlines headquarters in New York City.

This interview was very interesting: Apparently New York companies had experienced problems with new hires from outside of New York. The new employee was happy with his job, but his wife and family were not too thrilled with the congested living conditions. In too many cases the wives were unhappy and the new employee left shortly after moving to New York. So for my interview, Frank sent first class tickets (American Airlines, of course) for both Paula and me and put us up in the Waldorf-Astoria hotel—all first rate! We hired a baby sitter to care for Key and Cheryl and caught the American Airlines flight from Los Angeles to New York City in July 1967.

Paula busied herself with sightseeing while I met with Frank Kolk. As Vice President, Development Engineering, Frank was responsible for all airplane engineering for American Airlines. He had written the requirement for a wide-bodied, twin engine, fanjet airliner that would be much more economical to operate, much quieter, and much cleaner-burning than the existing fleet of Boeing 707s, Douglas DC-8s, and Convair 880s. His airplane was eventually produced as the three-engine (sorry, Frank!) MacDonnell-Douglas DC-10 and the Lockheed L-1011. It had three engines because of FAA's concerns for extended operations safety over water in the event of an engine failure. But Frank had a certificate stating that he was credited with inventing the L-1011.

Frank was a slightly built, frail man, having recently recovered from a mild heart attack. His Masters degree in aeronautical engineering from the Massachusetts Institute of Technology had qualified him imminently for his job at American Airlines, where he had worked since 1941. He had been instrumental in the development of the Douglas DC-4, DC-6, DC-7, and the DC-7C–perhaps the ultimate piston powered commercial airliner ever built. He had also ushered American into the jet age. Soft spoken, but he had a way of looking deep into your soul with his watery blue eyes that got your attention. Also, he knew all of the key aviation people in the airplane business, such as Donald W. Douglas (Douglas Aircraft), and James S. ("Mr. Mac") McDonnell.

Frank Kolk knew how to get what he wanted, and usually got it, one way or another.

I had lucked into a great reference book, printed by the Aircraft Industries Association (AIA), which was a lobby group representing aerospace airplane and engine manufacturers. I did some homework, and read up on the problems facing the airlines at that time.

After chatting informally for a time, I looked Frank in the eye and announced: "The way I see it, the problem is not how to get businessmen from LaGuardia to Washington National airports, but how to get them from their offices in Manhattan to their meetings in Washington D.C.. During the next five years, the airlines, collectively, will buy about $3.6 billion worth of airplanes, but the Northeast Corridor is so congested that you have no place to land them! STOL and VSTOL may be your best solution."

Frank blinked at me a moment in astonishment, looked down at the table, ran his fingers through his thin, slightly gray hair, then looked me in the eye and said: "You broke the code! How much money do you need to move to New York? You're 'top drawer!' "

That night, Paula and I had a wonderful dinner at a New York City restaurant. When we walked up to the door, we knew it would be expensive: there were chauffeured limousines parked at the curb, and the heavy wooden double doors were about twelve feet tall. There was a small, discrete bronze plaque next to the door announcing THE FORUM OF THE TWELVE CAESARS. We were properly dressed for the time and place in suit and tie, and cocktail dress. We were met by the Maitre D', impeccably attired in formal cutaway tuxedo and white tie, and escorted to a table set with silver, crystal, fine linens, and a small candle. Paula and I looked at each other in awe. It was a great place, and the service was impeccable. We knew it would be expensive! Fortunately, we were on American Airlines expense account, so we relaxed and enjoyed every luxurious minute of it. We had an appetizer, a nice bottle of French wine, a salad (Caesar, probably) rack of lamb, desert, an after dinner drink, and coffee.

It was the most elegant (and expensive) meal that we had ever had! With tip, the bill came to $35.00! (OK, but remember that this was in July of 1967, when a bottle of excellent French wine cost only about three dollars!)

We had seen many movies where the New Yorkers commuted to the city from Connecticut, so the next day we rented a car, picked a small Connecticut town at random, and visited a real estate office in Darien.

Little did we know at the time, that Darien, Connecticut, was the Beverly Hills of New York City! The real estate office had many pictures of houses for sale, but there were no prices listed on the pictures.

This was shortly after the Watts riots had enflamed Los Angeles, and we told the real estate agent that we wouldn't mind an integrated neighborhood, but we didn't want any racial problems.

He announced that Darien had no racial problems, because: "we don't let 'em in."

"You'll note that there are no prices listed for these houses," he explained. "You come in, for example, and that house is $45,000. If a Negro asks, then that house costs $150,000."

He showed us several houses. "That house right over there," he explained, "belongs to the president of the Coca Cola Bottling Company. And they are real nice people. We frequently see them our in the yard doing their own gardening."

We expressed our thanks, decided that this lily white and exclusive community was not for us, and returned to California to await a formal job offer, which Frank had assured me was forthcoming. Shortly before Christmas I got the offer of Development Engineer for VSTOL Technology at an annual salary of $18,500 (very nice for 1967) and asking when I could report for work.

I drafted a letter, explaining that we had to sell our house, and Key was still in school, and…and…and… and I realized that this was a dumb letter, and I wouldn't want to get one like it, so I tore it up and wrote that I could report for work the end of January.

This worked, and the date was set. I flew to New York City the end of January 1968, and stayed with my sister, Sally, who had a tiny apartment on the West Side just off Central Park. Sally was dancing on Broadway, and working at Continental Can Company, the same building at 633 Third Avenue that housed American Airlines Corporate Headquarters.

Getting Started—Assessing STOL and VSTOL State-Of-The-Art

On my first day Frank gave me my instructions: "I want American Airlines to be the leader in STOL and VSTOL technology among the airlines, I want you to figure out what your job is, and I want you to let me know what you are doing!"

WOW! What a terrific charge! I could do what ever I wanted! My success or failure would be on my own shoulders, unencumbered by someone else telling me what to do each day. Frank and I got along great! He was the best boss I ever had, and I sorely miss him. My first task was to fill out my knowledge of STOL and VSTOL technology relative to potential airline operations. I knew of all the U.S. and foreign programs (British, Canadian, German, French, Russian, and Japanese), but I needed to see, touch, and experience as many of them as I could first hand. I concentrated on the U.S. and Canadian programs.

I telephoned Helio Aircraft Company near Boston, Massachusetts. Helio had a small, single-engine, four-place aircraft with extreme short takeoff and landing capability. Walt Hodgson had test flown it at Edwards, and I had watched him actually hover over a spot on the Test Operations apron in a 22-knot headwind! The airplane had full span trailing edge flaps, spoilers for lateral control (small slats, called "fences," on the upper surface of the wing that could be raised or lowered to spoil the wing lift and provide roll control without yawing the airplane), and special slats (high lift devices) on the wing leading edge that would pop out automatically when air load changes warned that the wing was about to stall. As a result, the airplane could not stall— it would only start descending more rapidly. Helio disdained the so called "naked wing" STOL airplanes that relied only upon engine power and large wings for

Helio* Courier *STOL airplane

their STOL performance, such as Dehavilland's Twin Otter aircraft used by small, regional airlines, and the Swiss Pilatus Turbo Porter. "Unsafe," they claimed. I needed to sort this out.

Helio was more than happy to show American Airlines its capability, so I flew to Boston and Harry Wheeler, Helio's pilot, met me at the airport. He showed me the plant, and we went for a flight in a Helio Courier. He let me fly it a bit, but, not really being a pilot, I got more out of watching him fly. I asked him to demonstrate a short landing.

"OK." We were over a closed part of a runway near the edge of Boston's Logan Airport. We were about 500 feet above a runway intersection. Harry set the parking brake, pulled the throttle back to idle, and pulled the control stick back into his lap!

The little airplane "fluttered" down—that's the best description—at a high angle of attack. The leading edge slats popped in and out, and we set down gently in the center of the runway intersection. The wheels skipped twice, and we stopped in about two airplane lengths!

WOW! What fun!

[By the way: the little Helio Courier was still in production in 2002 — it can fly at 140 knots, take off and land in 300 feet, carry four people, and its minimum airspeed (it still does not stall) is 22 knots.]

Dehavilland, Canada, built a rugged, single engine "Bush" plane called the Otter, that Canadian and Alaskan bush pilots flew into remote, rough landing strips. Many of the Otter's had pontoons so they could land on the water when there was no other landing site. Dehavilland had increased the size of the Otter so that it could carry 19 passengers (the most passengers allowed under simplified FAA regulations for small airlines) and put two ultra-reliable Pratt & Whitney-Canada PT-6 turboprop engines on it. It was a simple airplane with fixed landing gear and no high lift devices other than trailing edge flaps. But it worked. It was inexpensive to buy and operate, and it was immediately snapped up, not only by Canadian and Alaskan bush pilots, but also by the "Third Level" feeder airlines in the U.S.. These were the very small airlines that served small communities and fed passengers to the larger airports for longer flights. Joe Fugere, President of New Haven, Connecticut-based *Pilgrim Airlines* had a simple rule for buying an airplane: If he could gross, with annual ticket sales, the purchase price of the airplane, he could make a profit with it. The Third Level airlines could do that with the Dehavilland Twin Otter.

***Dehavilland* Twin Otter**

So I telephoned Dehavilland Aircraft Company in Toronto, Canada, for a demonstration. Dehavilland readily agreed, and I went to the plant. Thus began a long and very enjoyable relationship with Dehavilland, and a wonderful associations with a very good friend, Bob McIntyre. Bob, a distinguished and cultured English gentleman, was Vice President of Marketing, and had a Master of Science degree in aeronautical engineering from Cambridge University in England. He also spoke fluent French, which, in Toronto, came in handy.

Hunter Blackwell, Dehavilland's Chief Test Pilot, briefed me on the Twin Otter and then took me for a ride. When I asked for a maximum STOL takeoff demonstration, he obliged, and it was most impressive! We leaped off the ground and hung on the props during climb out and turn. The Twin Otter was not really a STOL aircraft in the strictest sense of the word, but could easily fly passengers in and out of 2000-foot runways. Although too small for us, it was of great technical interest because of its STOL capabilities. More about Dehavilland in due time.

I also wanted to fly in Lockheed's rigid rotor helicopter, and, during one of my bi-weekly trips home to California, I arranged for "Fish" Salmon, Lockheed's crack test pilot, to pick me up at LAX airport and fly me to Van Nuys airport where Paula would meet me. "Fish" had flown first flights on many famous Lockheed aircraft, including the F-104 *StarFighter*. Paula's mother and father, having retired from the Air Force, were living in Van Nuys, where he was building the Santa Suzanna Rocket Test Facility for Rocketdyne. Rocketdyne was developing the awesome, 1.5 million pound thrust F-1 rocket engines for the huge Saturn V rocket that would eventually carry Neil Armstrong, Michael Collins, and Buzz Aldrin to the Moon on Apollo 11 on July 20, 1969.

Lockheed XH-51 Rigid Rotor Helicopter

"Fish" met me at the airport, and we had a great ride in the rigid rotor helicopter. "Rigid" had to do with the rotor hub, not with the rotor blades themselves. As a helicopter's rotor turns, the rotor blades are subjected to enormous stresses that could easily break the rotor hub. In order to eliminate these stresses, helicopter rotors were usually double hinged ("lead and lag" hinges—called "fully articulated") to the hub—they could swing up (all the way to vertical, in fact!), and down to a fixed stop that kept the blades from hitting the fuselage. And they could swing back and forth some. Centrifugal force held them out. This worked, but it also created great instabilities and difficulties in flying the beasts. Lockheed built a very strong rotor hub with no "lead and lag" hinges, which made the helicopter easier to fly, and also allowed a faster top speed. "Fish" demonstrated this to me before delivering me safely to Van Nuys and waiting Paula.

Fortunately, he did not demonstrate his flight (of which I have a film) of him flying under a low bridge over a dry river bed in the Santa Suzanna mountains!

During these first three months at American Airlines I commuted between New York and California and on alternate weekends I house-hunted in the New York Metropolitan area. I kept in close contact with Paula and the kids by telephone, and would catch the 6:00 PM American Airlines flight out of JFK Airport on Friday evening, arriving at LAX at 7:00 PM, where Paula would meet me and we would spend the night at her parent's house in Van Nuys. Sometimes I would take small gifts to the kids and sometimes I wouldn't—this to teach them not to always expect goodies every time daddy comes home. We would drive to Lancaster Saturday morning, and Paula would drive me back to LAX on Sunday to catch the noon AA Flight #21 back to New York. "Flight 21" was so named because its first class section featured meals from New York's famed "Club 21." With my position at American, I could fly unlimited space available–cross country in coach for $5.00, and first class for $8.00. That choice, of course, was a "no-brainer!"

In New York I would rent a car for the weekend and Sally and I would drive around New York (Up State, Hudson River Valley, and Rockland County across the river), New Jersey, and Connecticut, looking at houses.

I also flew in a number of other helicopters, including Sikorsky's S-61 and Boeing Vertol's twin rotor V-107 that were used to fly passengers from downtown (Los Angeles, Chicago, and New York) to their respective airports.

The Sikorsky S-61 was used in Los Angeles and Chicago. It was the most popular helicopter

Sikorsky S-61 Helicopter

because it was a single engine/single rotor machine that was more economical to operate than the Vertol machine. But its operating costs still had to be subsidized by the airlines. I flew in the S-61 from New York City to the Commons in Downtown Boston during a joint Sikorsky/Pan American Airlines demonstration of City-Center-to-City-Center operations. It was a vibrating, extremely noisy, uncomfortable flight.

The hostess provided by Pan Am got airsick.

Pan Am flew the Vertol V-107 from the top of New York City's Pan Am Building to Kennedy International Airport (JFK). The twin-engine, dual-tandem rotor machine was required because the New York City Port Authority demanded engine failure protection since it was flying over densely-populated New York City. It also had to be able to abort a takeoff, in the event of a single engine failure and land at its takeoff point. This meant that, when taking off from the top of the Pan Am Building, it had to takeoff backwards so that the pilot could see to land back on the building if an engine failed.

Boeing Vertol V-107 Helicopter

WOW! Now *that's* sporty!

I tried several times to take this flight to the airport, but never got the opportunity. The problem was that, even though New York City airport runways might be in clear weather and operating under Visual Flight Rules (VFR), the top of the tall Pan Am Building was frequently up in the clouds and under Instrument Flight Rules (IFR), under which the helicopter was prohibited from operating. The only flight I got in that helicopter was between JFK and LaGuardia airports—it was a vibrating, extremely noisy, uncomfortable flight.

During this time, I was also invited to give a day's lecture on VSTOL technology to the Aerospace Engineering faculty at the Air Force Academy in Colorado Springs, Colorado. I was invited by one of the Aeronautical Engineering Department's professors, Leo Stockham, whom I had met in the Edwards AFB Officer's Club Toastmasters Club. It was great fun, and he gave me a guided tour of the Air Force Academy—a beautiful campus with gorgeous Rocky Mountain views, and an incredible Chapel that was the pride of the Academy.

The Chapel, configured for Protestant, Catholic, and Jewish services, was a large A-Frame style structure with spectacular colored glass panels comprising the entire triangular wall behind the alter. Through these beautiful windows the congregation could enjoy grand views of the Rocky Mountains, so close you believed that you could reach out and touch them. The floor did not reach all the way to the walls—there was an open gap between the edges of the floor and the walls, which gave the impression that the entire floor was "floating" within the building.

Inside view of the U.S. Air Force Academy Chapel

Although pilot training is not provided at the Air Force Academy, students are given familiarization rides, but supersonic flights at low altitudes are prohibited since the supersonic shockwave of an early high speed demonstration had smashed out many of the expensive colored glass windows.

Thus I completed my initial assessment of STOL and VSTOL technology as it might apply to American Airlines short haul service. Basically, STOL seemed promising and worth further study, but VTOL was not viable: Helicopters were too slow, noisy, and uncomfortable, and too expensive to operate. Jet STOL technology was not advanced enough to evaluate, but it would have been too noisy for city center operations. Thus I concentrated my further studies on turboprop-powered STOL aircraft, of which there were two—the Dehavilland DHC-5 Buffalo, and the French Breguet 941.

The Big Move

In the early summer of 1968, Key had finished school for the year, and, even though we had not yet sold our house in Lancaster, I took a three-week's leave from American Airlines, and flew to California. We had a moving company pack our belonging, loaded the car, and Paula, Key, Cheryl and I drove through the South to visit friends, picked up my mother in Fort Worth, and drove to New York City. Mother stayed with Sally, and the rest of us took a hotel room for a few days while Paula found us a sublet apartment in Manhattan.

Many affluent New Yorkers sublet their downtown apartments, and spend the summer in the Hamptons at the Eastern end of Long Island, where it is cooler. There is South Hampton, West Hampton, and East Hampton. Their affluence determines which Hampton: it is said that the people in South Hampton talk only to those in West Hampton and East Hampton; that the people in West Hampton talk only to those in East Hampton; and the people in East Hampton talk only to God.

And there you have it.

The summer in Manhattan in our comfortable 15th floor apartment on the upper West Side gave us the opportunity to do all the "touristy" things, such as the Empire State Building, the Staten Island Ferry, and the Statue of Liberty.

The Empire State Building

When asked what they liked the most, Key and Cheryl both answered enthusiastically: "the ride on the subway!"

Paula found a beautiful, wooded lot in Northern New Jersey, that was within reasonable commuting distance, we bought it, and at the end of summer, rented a small house in Wayne, New Jersey, while we arranged to build our new house in the woods.

The Inflatable Airplane

Shortly after joining American Airlines, Frank sent me to a briefing at Rutgers University's Air Transportation Research Center. Rutgers is the New Jersey State University. The professor had scheduled a big press conference to announce plans for a new concept in short takeoff and landing aircraft. The presentation was presented in a wonderful old house that had been a safe house on the Underground Railroad during the 19th century. Many escaping slaves had sought and found safe refuge there during their flights to freedom in the North and to Canada.

The professor explained to the audience that the big problem with short takeoffs was the extra power needed. So he would augment the power needed only for cruise with "Accumulator Power" for takeoff. He would pump up a tank with compressed air, and use this compressed air to help power the engine for a short takeoff. the pilot would pump up the tank with the engine and then have about 20 seconds of extra power for his takeoff.

The press was taking notes.

He explained that the size of the tank would be a problem, so he would use the wings as the tanks. Of course, since metal wings would flex under the pressure they would eventually develop fatigue cracks. But not to worry! He wouldn't use metal, he would use rubber-impregnated fabric. To keep the wings from just blowing up like a balloon, he would connect the top and bottom surfaces with lines that would hold the wing's shape. He would pump the wing up to about 50 to 100 PSI pressure.

The press seemed fascinated by this concept, and was taking copious notes. I was also fascinated—*flabbergasted* may be a better word!

The professor continued: He could get even more power from the accumulator if he heated the air to 1500 degrees Fahrenheit!

That's when I looked over at the FAA representative, whom I knew, and she looked at me, and we tried not to laugh out loud!

The press had many questions, such as "When will these airplanes be operational?" "How big can they be?" "Who will build them?" and so on.

FAA and I were not impressed.

When I got back to the office, I reported to Frank: "He's a real nut case!"

"Yeah, I know." Frank smiled. "I thought you would get a kick out of it."

All in all, it was rather a fun day!

Dehavilland-Canada **Buffalo**—*Getting Over The Initial "Clutched-Up" Feeling*

My main interest in Dehavilland was in its twin turboprop-powered DHC-5 *Buffalo* military aircraft. The U.S. Army had bought a number of the earlier piston-engine DHC-4 *Caribou*, and had flown them in combat in Southeast Asia. The *Caribou* was not really a STOL aircraft—it had barely enough power for short takeoffs, and since it had no high lift devices, a short "assault" landing was a pretty "iffy" thing. The pilot would lower the flaps, set up the landing, reduce power, aim the aircraft at the landing point, and more or less just hang on and watch it slam down! The Army contracted with Dehavilland to build a better aircraft, which was the *Buffalo*. The *Buffalo* had two powerful GE-T64 turboprop engines (the same as in the XC-142A), high lift devices and flight controls, and was in every sense a high performance and controllable STOL aircraft. Dehavilland built six *Buffalo* for Army evaluation, and the Army was delighted with the new aircraft.

Then the Air Force and the Army got into a turf war over whose job was whose. The Air Force had better negotiators and came out of the meetings with responsibility for all of the Army's multi-engine, fixed wing aircraft missions. The Army retained its helicopters and single engine fixed wing aircraft. The Air Force brass—who had never flown an assault landing in a *Caribou* decreed that "we don't need no stinkin' new airplanes!" and canceled the *Buffalo* contract. After a few combat flights in the *Caribou*, however, the Air Force pilots announced "*Now* we know why the Army wanted a better assault aircraft!"

Too late—the Air Force brass would not admit its mistake, and the *Buffalo* was history! Except—Dehavilland was building them, anyway, and selling them to the Brazilian Air Force.

I had seen a *Buffalo* fly during a desert assault landing demonstration of our three XC-142A aircraft at Edwards, and was very, very impressed! It had positive control during both takeoffs and landings in very short distances.

About this time Eastern Airlines contracted with McDonnell-Douglas Aircraft Company, in St. Louis, to fly the French Breguet 941 STOL transport to simulate airline flight operations.

McDonnell-Douglas (MacDac) had the U.S. license rights to the 941, which they designated the MDC-188. Scott Crossfield had joined Eastern Airlines to investigate STOL aircraft for airline operations, and initiated this test program.

Scott Crossfield was the North American Aviation test pilot who had "invented" the X-15 rocket research aircraft. When he learned that Bell Aerospace was developing the powerful XLR99 rocket engine, he proposed to put it into a research plane in order to explore flight characteristics of atmospheric and exo-atmospheric designs capable of 4 to 6 times the speed of sound and 12 to 50 mile high altitudes. Scott subsequently flew the first flights on the X-15. Johnny Armstrong, no longer flying on flight test aircraft, had been assigned to flight planning for the Air Force, Army, Navy, and NASA flight tests of the X-15 as it explored the high speed flight regime.

Scott was arguably one of the most highly qualified experimental test pilots around: He had not only graduated from the Test Pilot School at Edwards, but also had a Masters degree in Aeronautical Engineering. When he finished his work at North American, he joined Eastern Airlines and headed Eastern's studies of STOL aircraft to alleviate air traffic congestion in the Northeast Corridor (Boston, New York, and Washington, D.C.). Basically, he was doing the same thing for Eastern that I was doing for American — except that he was looking at it from a pilot's view point, and I was looking at it from a total systems engineering viewpoint.

Frank Kolk decided that we should also evaluate the 941, and we contracted with MacDac for a similar flight evaluation program. More on this later.

In the meantime, I kept pestering Dehavilland for an evaluation of the *Buffalo*. Finally, one afternoon, I got a telephone call from Bob McIntyre. "We may have a *Buffalo* for you to fly towards the end of the month," he told me. "We don't have to deliver it to Brazil for a few weeks."

Thinking fast, I told him "That sounds great. I know that you're studying a four engine regional airplane based on *Buffalo* STOL technology, so I can see a big benefit from our airline pilot feedback to help you design your new Dash 7 airplane. Can we come up and fly the *Buffalo* for a week at no charge to American Airlines?"

"I'll check into it," he answered.

About an hour later he called back. "OK, " he said. "It's a deal."

Frank was in a meeting with Jack Graef, another engineer in our office, so I wrote Frank a short note:

"I seem to have arranged a one-week flight evaluation of a *Buffalo* STOL aircraft at Dehavilland, at no cost to American Airlines, but so far no one at American knows about this but me. We need to talk."

I handed this note to Frank. He looked at it, put it down, picked it up and read it again, and almost fell out of his chair! "Jeez!!" He exclaimed, running his hand through his thin, gray hair!

Frank loved stuff like this!

***Dehavilland DHC-5* Buffalo**

I wrote a 10-hour flight evaluation plan that would give me the information I needed to assess the *Buffalo*-type STOL aircraft and also give our two airline pilots a feel of how such an aircraft might operate in and out of congested airports, along Northeast Corridor airways outside of the regular airways. I had also included an hour for each pilot to get acquainted with the aircraft in order to ".. .get over the initial 'clutched-up' feeling of slow speed takeoffs and landings." This comment in the test plan caused great amusement to the Dehavilland folks but initial disdain to our pilots, who didn't really understand its purpose—until they actually flew a traffic pattern at 50 knots and came down to land on very short runways at twice the approach angles that they were used to! The Dehavilland test pilots understood completely, and commented "Here's someone who really understands the slow speed flight regime!"

The *Buffalo* flight evaluations were not all work: Toronto was a great place. It was scrupulously clean–especially compared to the trash-filled streets of New York City, and Paula brought the kids up for the touristy things. There were also great places to eat, which I enjoyed. When Frank came up and we started out to dinner, Frank had selected a place near the hotel, but I cautioned him: “You don’t want to eat there. I know a better place!”

This was an old mill converted into a wonderful restaurant, and the dinner was terrific!

Frank loved stuff like this!

Our flight evaluations went very well, and were mutually beneficial to both Dehavilland and to us. It was a fine program, and cemented the excellent cooperative relations that I maintained with Bob McIntyre and Dehavilland even after I left American Airlines. More on this later.

American Airlines' Inter-Metropolitan STOL Evaluation

I planned a much more detailed test plan for our evaluation of the Breguet 941/MDC 188 than for the *Buffalo* because we would be flying simulated airline routes and terminal area operations, and evaluating three new three-dimensional Area Navigation systems.

Normally, commercial airlines fly VOR routes between cities. The VOR system, for Very High Frequency, Omni-Directional Range, was composed of tall radio towers along FAA's defined airways. Pilots would dial a VOR station's radio frequency into their VOR receivers and a needle on their control panel would indicate whether they were flying right or left of the course—the object was to keep the needle centered. The pilots would dial in the radio frequency of the next VOR station when their instruments indicated that they had passed over the VOR tower.

Area Navigation, or "RNAV," would allow planes to fly in less congested airspace and save time and fuel. It would also enable more planes to fly the special RNAV routes than the VOR system could accommodate. The VOR system was "two-dimensional," meaning the planes followed latitude and longitude coordinates, following the VOR directions, but had to clear each altitude change with the FAA Air Traffic Controllers. The new three-dimensional, or 3-D RNAV, would enable them to clear their entire flight paths before takeoff, and set in their enroute directions and altitude changes as well. This type of routing would be crucial to realizing the benefits of short haul STOL service in the Northeast Corridor to relieve the crowded airways. We would be evaluating three different types of 3-D RNAV systems.

Scott Crossfield was completing Eastern Airlines' flight evaluation and had created a big flap with the Federal Aviation Administration (FAA). He had flown the aircraft, in Eastern's livery (airplane color scheme), in the Northeast Corridor using the British Decca Navigator, a 3-D RNAV system.

For this 3-D RNAV system he could program "waypoints" for his entire flight before takeoff, and then just follow the indicator needles—as long as he kept the needles centered laterally and longitudinally, we was right on course and altitude. He could obtain FAA clearance for an entire flight, and program the assigned altitudes of each waypoint. The system would automatically tune in the next waypoints as needed, and move the needles to indicate whether he should turn, climb, or descend. Since this new system did not rely on the raw VOR signals, which could vary depending upon obstructions or weather conditions, it was much more accurate than VOR navigation. The Decca system had a small computer that corrected signal anomalies, based upon LORAN C inputs, to yield this accuracy.

So far, so good.

But then Scott announced that the FAA's VOR system was no good, and would have to be replaced, based upon the discrepancies between his Decca system and the VOR system. Since his Decca system was so accurate, if the VOR system disagreed, then he claimed that the VOR system must obviously be wrong.

This annoyed the FAA. It didn't want to have to replace the entire U.S. VOR system.

My STOL evaluation team comprised me as Project Manager; our two pilots Bernie Wohl, STOL Project pilot, and Chuck Shafer; and Stan Seltzer, Director, Air Traffic Control Research. Additionally there were three representatives from American's Tulsa, Oklahoma, Maintenance Base: Bob Rogers, STOL Program Engineer; Jim Wilson, Supervisor, Maintenance Processes/STOL; and Jim Wheeler, Avionics Engineer/STOL. Joe Dobronski was MacDac's chief STOL pilot, and Jacques Jesberger, from Breguet Aviation would be there to teach our pilots how to fly at speeds below the power-off stall speed. There were also several Breguet maintenance personnel to maintain the aircraft. All in all, there were about 30 people assigned to or supporting my program.

We were in close communication with the STOL people at FAA, who were very interested in our work, for obvious reasons. I was coordinating with my engineering counterpart at FAA Headquarters (the lady engineer from the Inflatable Airplane briefing), Bernie was talking with his pilot counterpart, and Stan was talking about possible special airways routes with his counterpart. But we were getting different answers. Obviously, we needed a coordinated response from FAA. I was also concerned that, since we would be evaluating three different 3-D RNAV systems, compared with Scott's single system, that we would get involved in his flap with the

FAA over the VOR System. Something had to be done before we started our flights.

Frank was out of the country on a business trip, but I couldn't wait for his return, so I asked Stan to set up a meeting for me with Oscar Bakke, FAA Associate Administrator, who was a strong proponent for STOL service in the Northeast Corridor. In fact, in November 1966, he had turned an annual New York City emergency exercise for private and corporate aircraft into a demonstration of his ideas. STOL planes flew 550 operations out of Manhattan's piers and ball fields in two days. to simulate relief operations in the event of a large emergency. Stan set up the meeting, and I flew down to FAA Headquarters in Washington, D.C.

Oscar was very receptive to me. I explained that I needed two things from the FAA: First, our various contacts were very cooperative, but we needed a single voice from the FAA—a coordinated approach. And, second, I told Oscar that we would certainly get involved in FAA's flap with Scott Crossfield because we were doing similar tests, but that we had no preconceived opinions about the validity of the FAA's VOR navigation system. I told Oscar that we needed a navigation yardstick, to measure our navigation accuracy, that was more accurate than the claimed accuracies of the three 3-D RNAV systems that we were evaluating.

He listened attentively, told me that he would let me know, and I flew back to New York. The next morning, my engineer contact telephoned me. She told me "Oscar had a big meeting with us this morning and there is no doubt in anyone's mind here that (1) we are fully supporting American's STOL evaluation program, (2) that Bob Meyersburg (FAA's STOL pilot) is in charge here, and (3) that we will set up a highly accurate tracking radar in Chicago to monitor and track your flight paths to compare against your aircraft measurements!"

This was exactly what I wanted to hear!

About this time, Frank returned, so I briefed him on what I had done. He listened in amazement, then put his head down on his desk between his hands, ran his hands through his thin hair, and exclaimed "Jeez! You went in there fully loaded! Oscar didn't stand a chance!"

Frank loved stuff like this!

We had decided to fly most of our evaluations from Chicago, with a short final series of flights from LaGuardia. But I needed to decide whether to fly the Chicago flights from O'Hare (the busiest airport in the country) or Midway Airport (under utilized, in a noise-sensitive neighborhood, but very convenient to downtown Chicago). I talked to everyone but got no definitive information. So I announced to everyone that we would base the flights out of O'Hare–sat back, and waited for it to hit the fan! Within two hours I had all the information I needed to know that O'Hare was a really, really dumb idea, and that the flights should be based from Midway! I made it so.

In the meantime, MacDac was painting the airplane in American's new livery (the wide red, white, and blue side stripes and the stylized eagle on the tail—it was the first plane to have American's new livery). MacDac was also installing a special photo panel to record our navigation data, and the three 3-D RNAV systems. Each of the three systems used a different technology so it would be a good test of the various approaches. The navigation companies were providing the equipment and support at no charge in return for the flight test information that we would develop and provide to them. Our pilots would first fly training and familiarization flights out of MacDac's St. Louis facility, which shared runways with St. Louis's Lambert Field commercial airport.

Breguet 941/MDC-188 STOL Transport

The contingent of French pilots and maintenance personnel would actually take care of the airplane and fly with our pilots. Fortunately, our maintenance officer, who was evaluating the maintenance characteristics of this type of airplane from an airline standpoint, spoke excellent French, and could read the French maintenance manuals in French. He had been with Trans World Airlines in Paris for ten years. The most unusual characteristic of this airplane, from a maintenance standpoint, was that it had a cross shaft connecting all four engines. Since the wing depended upon propeller slipstream to keep from stalling at its slow takeoff and landing speeds, the cross shaft prevented loss of a propeller prop wash over the wing in the event of an engine failure. This was essential when flying below the power-off stall speed. When the first engine was started, all four propellers began to turn. We had learned that this was critical for flight safety years before at Edwards with the X-18, which did not have cross-shafting.

I was flying to St. Louis on Monday mornings, spending the week in St. Louis, and flying home on Thursday night to be in my office all day Friday. Frank flew to St. Louis to attend one of our planning meetings. During the meeting, the MacDac people asked a question and looked to Frank for an answer–after all, he was a vice-president.

Frank just sat there, ignoring them, then finally looked up and answered "Don't ask me, ask Ransone, he's running this program!"

What a guy! After that, there was never any question of who was in charge, and I got everything I needed.

I met “Mr. Mac.”

Mr. Mac, formally known as James S. McDonnell, founded McDonnell Aircraft Company in 1939. He built a remarkable company that designed and built fine airplanes and treated its people right. Unusual for aircraft companies at the time, MacDac employees had their first names on their badges, and James S. had "Mr. Mac" on his, and that is what he liked to be called. One Friday afternoon Marvin Marks, MacDac's STOL Program Manager, announced that Mr. Mac wanted to see me before I flew back to New York. "Mr. Mac talks very slowly," Frank Kolk had warned me, "in fact, he talks so slowly that you'll think he’s died."

About 4:00 o'clock, Marvin escorted me to Mr. Mac's third floor office—a huge, mahogany-paneled and furnished enclave that took up an entire corner of the building. Marvin introduced us, and left. Mr. Mac was about 70 years old, was tall, with thin gray hair, and a friendly expression on his face. Impeccably attired in a business suit he moved slowly, but stood up and walked around his huge desk to greet me cordially and ask if I wanted anything to drink. Then we chatted. It was not the conversation that I had expected. He never asked me about our plans for buying STOL airplanes, but asked about me, my family, and my career. Sure enough, he talked so slowly that at times I was certain that he had died and just not yet toppled over. It was getting close to time for me to catch my plane back to New York, and I was just on the point of telling him that I could easily catch a later plane—after all, it would depart from the terminal just across the runway from his office. But he was well aware of the time.

"I understand........................... that you.......................... have.. [I think he just died..].............a 5:30 plane."

"Yes sir," I answered, "but I can..."

"My car.........................is at the front door,............................... ..[yes, I think he's died] ....and it will...................... .take you across the runways to........................ .... [OK, he's gone this time, for sure!].............your plane."

What a gentleman! It was easy to see how MacDac was such a successful aircraft company. I will always remember my one meeting with "Mr. Mac."

I had added 20% contingency funds to the evaluation budget for unexpected problems or opportunities. This came in very handy. Bob Rogers and Bernie Wohl had drafted the detailed flight program in the Chicago area to evaluate not only the terminal area and routes of interest, but also to evaluate the accuracy of the three onboard RNAV systems. Remember that the FAA was tracking us with its precision radar so we always knew exactly where we were—or at least where we had been. Our location was measured in terms of rho-theta and altitude. Rho is the distance from a known reference point and theta is the angle from some known datum direction to that reference. Bob had defined a large matrix combination of rho and theta, but I was looking for some way to simplify and reduce the flight hours to save time and money.

I used some of my contingency funds to contract with Battelle Memorial Institute, a Columbus, Ohio, "Think Tank" to study this problem. Battelle had gotten its financial start several years before when, as a small, non-profit company that supported entrepreneurs, it was approached by a fellow who had a new idea on how to reproduce documents. Battelle liked the concept, funded additional research, and bought the Haloid Copy Company to manufacture and market the new duplication system. They renamed the new company "Xerox."

I focused on the rho-theta combinations.

"The distance measurement error will probably vary with distance from the reference point, won't it?" I asked. Jim Loomis, of Battelle, agreed. "But won't the theta angular measurement be the same regardless of distance from the datum?" Jim scratched his head and thought about this.

"By golly, I think you're right!"

This enabled us to cut our test time almost in half because we did not have to test at all of the rho-theta combinations originally planned.

Never be afraid to ask the "stupid" questions.

After our pilots had finished their familiarization flights in St. Louis, it was time to move our operations to Midway Airport in Chicago. American's Public Relations Department, always on the lookout for a publicity event, arranged for Chicago's Six O'clock News team to be on hand for the MDC-188's first arrival at Midway.

Now you need to know that the City of Chicago had been trying to get the various airlines to expand their operations out of Midway in order to relieve some of the congestion at O'Hare. Unfortunately for this plan, however, travelers didn't want to spend an extra hour commuting between O'Hare and Midway to catch a connecting flight. Consequently, the Midway Airport was very underutilized, and the surrounding area was economically depressed. Stores were empty, the terminals were mostly deserted, and there were no airport concessions. Just before the plane arrived, I was interviewed by a local television news anchor. He was smartly dressed and sported a pancake makeup suntan (so on screen I probably looked like I had died the night before). The camera was set up, he got in position, and I was prepared for his first question about this amazing new aeronautical technology.

"I have here Mr. Rob Ransone of American Airlines, who is conducting flight tests of a new type of short haul airplane. Mr. Ransone, will this new airplane revitalize Midway Airport?"

WHAM!

I wasn't expecting a question like that! Thank goodness for Toastmasters training to think on my feet! I simply reoriented the question to one that I could answer without getting American Airlines in political hot water.

"Whether Midway Airport will be revitalized depends on a lot of things, but as far as this aircraft is concerned, the runways are plenty long and the airplane's quiet sound levels will not disturb the airport neighbors!"

Rob—you silver-tongued, glib, devil—you've done it again! Fortunately the airplane arrived before any more difficult questions were asked, and the cameraman focused on the arriving plane.

American Airlines' Midway Airport STOLport

We wanted to fly into Merrill C. Meigs Field near downtown Chicago, in order to demonstrate operations from small city-center runways but had to get permission from Chicago's crusty old Commissioner of Aviation, Bill Levine (I'm not sure this is really his last name). Bill was a National Football League (NFL) umpire, and took no gruff from anyone—a great guy to have on your side. Bernie knew Bill very well, and introduced me. Bernie explained that we wanted to fly in and out of Meigs Field, and needed Bill's approval because the Meigs Field airport manager had refused.

Bill explained that the approach path to landing at Meigs Field was headed directly into the city center, so that if the plane missed the turn on final, it would fly into the downtown buildings. He told us that he would argue with *our* lawyers, but that he would not argue with *his* lawyers.

Bernie explained the navigation equipment we had on board, our slow approach speeds, and promised that we would fly only in clear weather. Bill listened attentively, asked a few questions, then picked up the telephone and called the Meigs airport manager. His side of the conversation went something like: "Hi. This is Bill. I've got two American Airlines people here who want to fly into Meigs for their tests. They know what they're doing, and I think it would be OK. What do you think?..........................*I said, I think it will be OK—what do you think*?..........OK, great. Thank you."

He hung up and turned to Bernie. "OK. Anything else?"

Bernie told him that we wanted to take off and land from the O'Hare airport taxiways during active runway operations, and explained that, because of our slow takeoff and landing speeds, and slow speed and low altitude maneuvering, it would not interfere with normal airport operations. We needed this information to show the capability of this technology to alleviate airport congestion. Bill called the O'Hare airport manager. His side of the conversation went something like: "Hi. This is Bill. I've got two American Airlines people here who want to fly from the taxiways at O'Hare for their tests. They know what they're doing, and I think it would be OK. What do you think?..........................*I said, I think it will be OK—what do you think*?..........OK, great. Thank you."

After arranging everything we needed, Bill got a call from his boss (then) Chicago Mayor Richard J. Daley— when Bill called his airport managers *they* jumped, but when His Honor Richard J. Daley called Bill, *Bill* jumped! We thanked Bill for his help, and took our leave.

When I got back to Midway, one of Chicago's Finest was sitting in my office. I introduced myself and asked what I could do for him. He sat back in his chair, eyeing me coldly, and flipping his billy club. "I want a ride in that airplane," he announced, smugly, and waited for my response.

I had heard all about the Chicago police and how tough they could be, but he didn't know that I was now a good friend of Airport Commissioner Bill Levine! I looked him squarely in the eye and told him: "That's great! I'm delighted that you are interested in what we are doing!" I then explained our program, and told him that we had limited seating and that it was full of officials from Chicago, Federal Aviation Administration, the U.S. Department of Transportation, The National Aeronautics and Space Administration, and various airlines and navigation system companies. I finished up with ".. .so we don't have room for you right now, but come back in about three weeks and I'll get you on the plane! Thank you so much for your interest. Keep up the good work!"

With that, I escorted him to the door, breathed a sigh of relief—and never saw him again!

I was still working with our test team during the week and flying home to be in the office every Friday. One Friday I discovered that American Airlines' president, George A. Spater, would be flying to South Bend, Indiana, for a meeting the following week. I sent him a note, inviting him to fly into Chicago and we would fly him to South Bend in the MDC-188. His secretary called me and announced that he would like to do that.

The following Tuesday morning George Spater and Frank Kolk arrived at our offices at Midway Airport. Warner Lowe, the MacDac STOL Transport Demonstration Manager, briefed Spater on the airplane, I briefed him on our STOL Evaluation Program, and Bernie Wohl briefed him on today's flight. I had emphasized that, since these flights were costing American Airlines

Breguet 941/MDC-188 at American Airlines' Midway Airport STOLport

$4,500 a flight hour, that this was not just a joy ride—we would be following a specific flight plan and collecting test data all the way.

Spater liked this.

Frank loved it!

The flight went well, we arrived on time for our Public Relations folks to cover the event with the South Bend news people, and our arrival was on all of the 6:00 o'clock television news channels.

Everyone was happy.

But one time I did something that didn't make everyone happy. Bob Rogers telephoned me during my Friday in New York that Bernie Wohl wanted to fly the Breguet into Butler Field, just northwest of Chicago. Paul Butler was a good friend of Bernie's and had made one of the 3D-RNAV systems we were evaluating. Fine.

However: there were some safety issues that bothered me. First, the Breguet had a very narrow landing gear configuration and a high center of gravity. Both contributed to a higher "tip-over" tendency than most airplanes. Second, because of the high engine and propeller configuration and the way the propellers generated thrust, cross winds could aggravate the problem. Third, during a French flight test, the airplane had actually tipped over, an outboard propeller had struck the ground, shattered, and some flying pieces had pierced the fuselage and injured a crew member. Finally, it had been raining all week, and Butler Field's grass runway was wet and soggy.

All of this really bothered me, so I told Bob that, because of these things, I thought the senior French test pilot should make the landing at Butler Field. Bob told Bernie.

On Sunday night, when I arrived back in Chicago, Bob told me that Bernie was very, very upset with me. Bernie had wanted to land the plane and show off to his friend Paul Butler. The French pilot had made the landings, with Bernie sitting in the jump seat behind him, with his arms folded, refusing to do or say anything, and the French pilot exclaiming "Ou la-la! Ou la-la!" over and over. Not only that, but Bernie, Chuck, Bob, and Warner Lowe had brought their wives along! This on a test airplane, landing for the first time on a wet, grassy field, with dissension in the cockpit! Not a good thing!

I swallowed my pride, admitted to a failure in judgment and smoothed everything out. My bosses at American, Frank and his boss George Warde, were sympathetic and only said "Well, you won't make that mistake again!"

It would have been a different story if the airplane had hit a soft spot on the field and tipped over–then it would have been the French pilot and his company who would have been to blame, not Bernie or American Airlines. I may not have made a *popular* decision, but I still think it was the *right* decision. My big mistake, however, was in not telling Bernie that my decision was based solely on the liability issue, not on his ability as a pilot.

Both NASA Ames and NASA Langley were doing valuable research on STOL and VSTOL. Langley was concentrating on VTOL, and Ames was concentrating on STOL. In fact, Ames had test flown the Breguet 940, and earlier version of the Breguet 941, in France. I thought it would be a good thing for the Ames folks to fly the 941, so I invited them to Chicago, and they readily agreed.

Bob Innis was the test pilot, and Seth Anderson and Kurt Holzhouser were the flight test engineers. They spent two days with us and had two very productive flights. We were also collecting our own flight test data, so we didn't lose any test time, and their comments were very helpful to us because we could then relate the 941's performance and handling qualities to other STOL concepts. It was also very useful to Ames because they could evaluate the 941's improvements over the 940.

The last week of our testing, before moving operations to New York's LaGuardia Airport, I wanted to show our appreciation to our test team and Chicago hosts by having them to dinner. Frank agreed, I found a very nice restaurant and made reservations for about 30 people. We had an open bar and each diner had a choice among beef, chicken, or fish.

Paula and Frank flew in from New York, Warner Lowe and his boss Marvin Marks were there, as was American's Chicago Operations Manager. The afternoon of the dinner, I told Jim Wilson that I wanted to make a little speech in French. Since Jim spoke fluent French he could teach me what to say. I told him that I wanted to tell the French that we enjoyed working with them, we appreciated their support, they had a fine airplane, and we wished them luck.

Jim thought about this a few minutes, and then taught me what to say. I practiced it over and over and started feeling pretty proud of myself. Then I had a sudden thought:

"When I make this speech in French," I asked Jim, "do you think the French will be upset and think that I could speak French all along and was just fooling with them?"

Jim laughed. He actually *laughed*: "No chance!" He replied. Not "I don't think so," or "probably not," or some other gentle statement. But "*No chance*!" I was crushed!

That night we had an unexpected visitor from France: Mess. Henry Ziegler, President-Director General, De Sud-Aviation, Societe Nationale de Constructions Aeronautiques, 37, 3d de Montmorency, Paris, XVI0. Sud Aviation was the parent company of Breguet Aviation, and Henri Ziegler was the father of the Breguet 941 and also of the Anglo-French *Concorde* supersonic airliner. A very impressive guest.

After dinner, I made my speech to the gathering, and finished with my carefully practiced French. And it was the hit of the evening!

I got a rousing, standing ovation from the Frenchmen! Not for my fluency in French, nor my Texas-pronunciation of pidgin-French, but for the effort that it obviously took for me to pull this off.

Rob, you glib, silver-tongued devil—you've done it yet again!

After we completed our flights in the Chicago area, we moved our operations to New York's LaGuardia Airport for terminal area evaluation in the congested Northeast Corridor and STOL approaches to a simulated Hudson River STOLport. These were great fun. We could fly north up the Hudson river at about 500 feet altitude and 50 knots, make a u-turn within the confines of the river at midtown Manhattan, and descend to about 100 feet along the Hudson River waterfront. We were allowed to do stuff like this because of our close working relations with the FAA and the New York City aviation authorities.

I had used some of my contingency funds to have a photographer take 16mm movies of our flights in Chicago and New York, and took some myself. I bought a film editor, and spliced together the original film to make a draft filmed report of our evaluation. I wrote the script, and had our Audio-Visual Department in Tulsa, Oklahoma, put together a 25-minute film. I gave copies to the FAA, New York Port Authority, MacDac, Breguet, Battelle, and NASA as well as copies throughout American Airlines. During the following year it was shown over 60 times in six countries to more than 2000 interested people.

Hudson River from Breguet 941 cockpit

We had proved that STOL aircraft could, indeed, fly safely in otherwise unused airport and airways space to relieve congestion. We provided high quality performance data on three 3D-RNAV systems. We saved the FAA's VOR System. We updated NASA's STOL performance database. Jim Wilson determined that the only maintenance difference between this STOL aircraft and conventional aircraft was the cross-shafting between the engines/propellers, and that this complication had about the same maintenance costs as another propeller.

And I completed the evaluation $638 under budget! Frank said that no one had ever done such a thing at American before.

What fun!

And Frank loved it!

A Carefully Planned and Executed Political Demonstration

One day Grumman invited me to join them for a lunch at which the Manhattan Planning Commission was presenting its Five Year Plan for the Borough. Grumman had bought a "table" for ten people, and they invited me to join them because of my work with short haul air service.

When we got to the hotel conference room, I noticed a table in front of the door with bullhorns, TV camcorders, microphones, etc., but thought nothing of it.

After we all had sat down at our table—there were about 20 tables or so—I noticed that the TV news cameramen were all gathered at the front doors, which were closed. On a signal, the TV lights came on, the cameras started rolling, and the doors burst open and in rushed about six people! After they were in, the cameramen stopped their cameras and turned off their lights. While we were eating, the demonstrators came to each able and quietly passed around printed flyers that expressed their views about what they believed was important to the Five Year Plan.

After we finished our lunch, the speakers at the head table began their general presentation, which would be followed afterwards by individual group meetings in separate rooms. As they started their presentations, the demonstrators began shouting their demands. They jumped up onto the head table and snatched the microphone from the speaker. The speaker turned the microphone off, but the demonstrators had their own battery-powered bullhorns, through which they shouted their demands.

Finally the planning meeting organizers just cancelled the luncheon, and told everyone to go to the individual meeting rooms. We didn't go, but I understood that the demonstrators attended these individual meetings, and, therein, out of camera range, quietly and cooperatively expressed their concerns and wishes.

I called Paula and asked her to watch the Six O'clock News. She did, and when I got home, I asked her how many demonstrators there were. She said there must have been a lot, because they filled the TV screen. I told her that there were only six, but the way the news media covered it, it seemed like a large mob.

The whole thing was carefully planned and well organized and coordinated with the media.

Controversy is always more newsworthy than cooperation, and always sells more newspapers and television time. Years later, as a Visiting Associate Professor at The University of Virginia, I taught my graduate students that the *New York Times*, the *Washington Post*, the *Los Angeles Times*, and the *Dallas Morning News*—as well as all of the television news networks—do not sell news, they sell newspapers and air time, and controversy is more salable for advertising revenues than cooperation.

Floating Interim Manhattan STOLport - FIMS

The FIMS story is a bit lengthy, but there are so many lessons to be learned from it that I believe it is worth relating in considerable detail. Specifically, it makes the point that engineers must be aware of the socio-political, economic, and environmental issues involved in new technologies implementations, as well as the technical and engineering requirements—systems engineering at its broadest definition.

Engineers don't learn these issues in school, but they can stop any technology implementation in its tracks.

As our STOL evaluation drew to a close, and we were finalizing our written and film reports, I started thinking about the next phase. Now that we understood the STOL airplane and navigation technology state-of-the-art, what was the next logical step? Obviously, we needed to demonstrate short haul STOL's capabilities and benefits and identify any specific problems still to be resolved. FAA Air Traffic Control system changes, aircraft certification, and perhaps even financial support, indicated that there should be a bona fide, fare-paying demonstration, and that this demonstration should be between Manhattan Island and Washington, D.C.

Bob Meyersburg flew me around the Washington area in the FAA's ancient DC-3, and we found a few potential STOLport locations that would be convenient to the District. We also thought that we could find suitable STOLport sites in the metropolitan Boston area. But Manhattan would be the tough one, and without a Manhattan STOLport there could be no Northeast Corridor STOL service demonstration.

There was certainly no place on Manhattan Island to land, so I concentrated on the rivers. Airplanes operate from Navy aircraft carriers all the time, so why not build a floating STOLport? I made some very rough estimates of what I thought 2000 feet of iron barges, with a landing strip on top, might cost. Based upon 1/4-inch steal plate at $100/pound, and some other wild guesses, I estimated a construction price of about $12 million. I discussed this with Frank. He thought this was a terrible idea, and ran me out of his office.

So, I just waited, and, sure enough, the next day he came into my office and announced that, on reconsidering, he thought it was a great idea. It was not the only time that Frank tossed me out of his office because of my wild ideas, but he always came back and agreed.

What a guy!

With Frank's approval, I asked Bob McIntyre, at Dehavilland, to have its graphics department draw up an artist's conception of a floating STOLport, and sent him a photograph of the Manhattan waterfront where it might be located. They did a wonderful job, which helped me to sell the idea.

At any rate, we convinced American Airlines to bite the big bullet and study a STOL service demonstration for the Manhattan-Washington, D. C. route. Furthermore, a Hudson River STOLport site between west 26th and west 34th streets near the community of Chelsea, would be the study site for the Manhattan STOLport because the approach paths at this area would parallel the shoreline and not require heading towards the city.. The presence of an already organized citizens group against STOLports at Chelsea was considered an asset. I knew that such a group would inevitably form wherever we located. Since this one already existed, I thought we could discuss our plans with them in detail.

A STOL service demonstration's success, measured in terms of how well it answered the many questions facing the airlines regarding market acceptance and economic viability, would depend upon many elements. These were the STOL aircraft, their certification by the FAA to ensure their flight safety but not legislate away their slow speed performance advantages; STOLports convenient to travelers and their destinations; Air Traffic Control procedures that would enable their unique flight path and air route benefits to be exploited; and the latest technology guidance and navigation systems. I initiated study programs in each of these areas.

We needed six STOL aircraft that could be certificated by the FAA for fare-paying passengers. The Breguet 941S would have been much too expensive to modify for FAA certification and it was not in production. The only aircraft possible was the Dehavilland Canada DHC-5A *Buffalo*, which was about the same size as the Breguet 941S. The *Buffalo* was still in production, and had been provisionally FAA certificated during its development for the U. S. Army. In May 1970 I sent a Request For Proposal to Dehavilland for six *Buffalo* aircraft, outfitted to airline requirements, and certificated to operate safely from 2000-foot STOLports. These would carry 36 passengers.

In our discussions, the FAA had promised us special airport procedures and air routes that would maximize the benefits of STOL aircraft capabilities to relieve congestion. Now I went to the FAA and announced that we had sent a specification to Dehavilland for six airplanes, and asked them point blank: "...where are the specific procedures and air routes that you promised?"

"Uh, uh, well, uh...but...but...but...." The FAA had been greatly supportive when speaking in general terms, but now that we needed specifics, they suddenly found fourteen (yes, I counted them) reasons why it couldn't be done. Fortunately, I had strong support from FAA technical people, the U.S. Department of Transportation, NASA, various aircraft and engine manufacturers and their lobby groups, and other airlines. Now that I had their attention, our people could work with them to solve these fourteen specific problems. They started to work out solutions.

The Manhattan STOLport

Concurrently, I focused my attention upon the Manhattan STOLport. For this requirement, I conceived the idea of a simple, relatively inexpensive, floating STOLport, which we felt could be located in the Hudson River adjacent to Chelsea, but towed to other locations as necessary for other site evaluations. Furthermore, it did not require extensive foundations or footings, could be easily modified as required, and could be sold for scrap when no longer needed. I called this the Floating Interim Manhattan STOLport, or "FIMS", and released the Request For Proposal to several architect and engineer firms. I evaluated the three responses, and selected Howard, Needles, Tammen and Bergendoff to do a $36,000 Technical Feasibility Study.

Unfortunately, American Airlines had to economize because of a moderate business recession, and felt that it could not afford the study at that time. So I set up another meeting with Oscar Bakke and flew to Washington to convince him that the FAA should fund this study. I pitched my case and flew back to New York.

The following week, I telephoned my engineering contact at FAA to inquire whether I should come down to explain anything. She told me that the ways were greased so well to approve this, that I could only screw things up! This was in early September, the end of the FAA's fiscal year, and FAA had fallout money they had to spend. Within two weeks from my presentation we had a signed purchase order from the FAA for the $36,000 study. I got American to sign the agreement in only one week by threatening that we should be twice as fast as the FAA! We signed the study contract with Howard, Needles, Tammen and Bergendoff, and planned to just sit back and await the report.

Not so fast! A government contract requires public notification, by law. The nature of this notification was considered. The question of course was which of the various groups should be notified in advance of the news release.

On June 24th, I telephoned various city officials in the NYC Department of Transportation, the City Planning Commission, the Department of Marine and Aviation, and the Port of New York Authority (recently renamed The New York and New Jersey Port Authority, in order to acknowledge the fact that New Jersey shared the Hudson River), and told them of the FIMS study. I also considered contacting the Chelsea Against the STOLport Committee, but decided not to. I felt that we would get acquainted in due course, but that such a contact at that point would imply American Airlines' acceptance that this committee indeed represented the citizens of Chelsea. I felt that there may be others in the community who would be in favor of FIMS because of its temporary nature as opposed to the very permanent characteristics of the earlier concepts, and I wanted these other groups to have an equal chance to step forward. In hindsight, this was probably a bad idea. I should have telephoned their chairwoman, Ms. Jacqueline Schwartzman, and told her of the study as a matter of courtesy. American Airlines prepared the news release, it was approved by the FAA, and duly appeared in the July 2, 1970 New York Times.

That's when it "hit the fan!"

The Manhattan STOLport

Now you need to understand a bit about New York City politics: As John Wiley, Director of Aviation for the New York and New Jersey

Port Authority commented: "From the top of the Empire State Building I can see 2500 different political entities! Any one of these *can stop* something, but you have to have almost *all* of them in agreement to actually *do anything*." Manhattan residents know this full well, and had organized "Block Associations." A Block Association typically comprises the residents on both sides of a street for one or more city blocks—say, the 300 Block West 22nd Street Association. This gives the residents enough voting clout to get the attention of the Manhattan politicians.

A year or so before, Ben Darden, consulting for the New York City Department of Planning, had proposed a huge office building for the West Side along the Hudson River. The unusual thing about this building was that it did not stand up straight, like most dignified Manhattan Skyscrapers, but was reclining languidly on its side! It would have been 800 feet wide and 2000 feet long to accommodate a STOL runway, and would have been ten stories tall to get the airplanes above Hudson river shipping. At 16,000,000 square feet, this would have been the world's largest office building! Unfortunately, this site would have been next to an affluent neighborhood known as Chelsea. This was a close-knit community of professionals, artists, and business executives. They had organized a large number of these Block Associations, and soundly defeated Ben's audacious plan for a STOLport.

Now, I had unwittingly kicked open this hornet's nest with my floating STOLport idea! The Chelsea reaction was immediate. The Chelsea Against the STOLport Committee held a press conference July 10th next to the existing 30th Street Heliport, at which they denounced the STOLport study. Several of the Chelsea political representatives were there, and a statement by U. S. Congressman Edward Koch, 17th District (NYC) was read.

I thought that if Congressman Koch knew the details of our study, its purpose, and that it was only one part of a much larger effort, he could assure the residents of Chelsea that there was nothing to fear, and we could discuss their concerns. How naïve? Therefore I telephoned him in Washington, and tried to set up a meeting with him. He refused. He said he would not see me, but I could send him some material. He was against the STOLport and against the study. I sent him some information, which he did not acknowledge.

I was amazed: How could he represent the people if he refused to get all of the facts? To better understand his position, first I had to assume that he was acting exactly right from his viewpoint. Next, I hypothesized some possible reasons for this action. These may not be his, but they were ones that might have prompted his reaction, and helped me to understand and to accept his position. First, if his constituency were so vehemently against the STOLport, the last thing a smart politician would want to do would be to get the facts, tell them they were mistaken, and that the STOLport was a good idea. Furthermore, if my arguments were convincing, he might have difficulty supporting his constituents' position. It would be better to avoid any contact with me. Unfortunately, this position closed all channels of communication between us, and we had no further personal contacts.

First Confrontation

The first direct confrontation between American Airlines and the Chelsea Against the STOLport Committee occurred July 29th, 1970, at P.S. 11 in Chelsea. I represented American Airlines and Dave Lobb of our Public Relations office was there to keep an eye on our image, and on me, no doubt. Our speaker was James Pyle, Director of the Aviation Development Council. He was based at LaGuardia airport, and represented all of the airlines serving the New York area and also the Port Authority of New York and New Jersey.

Mr. Pyle was the first speaker, and carefully explained the purpose of the study. He emphasized that this was not a study to *build* FIMS, but only a technical feasible study to *obtain information*. The audience was attentive, courteous, and asked the questions any of us would ask in similar circumstances. I was encouraged. Of the 300 or so residents in attendance, most seemed *not* to be strongly opposed to the FIMS study, although a few were very strongly against the study as well as the concept. I felt that they would probably prefer not to disturb the status quo, however, and would feel more comfortable without the FIMS study.

Oh, did I mention that this was also election time for a number of the politicians? There had not been an issue to attack until this, but *now there was,* and every politician pounced upon it with great, pontificating delight!

It was quite a show! The speeches addressed the Vietnam war, the U. S. supersonic transport, motherhood, religion, care for the aged, air and noise pollution, safety, and Big Business Arrogance. Congressman Edward Koch was there. Congresswoman Bella Abzug was there under one of her huge, trademark hats. Every political hopeful was there milking the issue for all it was worth. This is not a criticism of them or politicians in general; it is mentioned to show the situation in which American Airlines, and I personally, now found ourselves. The most interesting speaker was a lady judge, about 75 years old. A wonderfully impressive speaker. Afterwards, I met her and told here that, although I disagreed with everything she said, that I sure admired the way she said it!

"If you think *that* was good," she replied, with a twinkle in her steely gray eyes and a tiny smile lurking at the corners of her mouth, "wait until I'm on *your* side!"

What fun!

We were treated courteously by the residents, and especially by the Chelsea Against the STOLport Committee and their chairwoman, Ms. Jacqueline Schwartzman. Jackie was the wife of a Chelsea dentist, and worked part time in his office. But she still had time to pull together 44 neighborhood block associations into the Chelsea Against the STOLport Committee. We met her and chatted briefly, explaining that we were not actually *building* a STOLport, only *studying* the concept.

Jackie was not mollified. She had had dealings with big corporations and city officials before, and knew that she had to start early before things got out of control in order to have any chance of stopping this encroachment on her precious community.

All in all, we were treated cordially, considering the enthusiasm of the other attendees, and I was relieved to survive without tar and feathers. The most productive outcome was that Jackie and I met, and I soon established a close—albeit controversial—relationship with her and her committee.

Activities of Both Sides

From August through September, as Howard, Needles, Tammen and Bergendoff were conducting our FIMS feasibility study, the Chelsea Against the STOLport Committee was very busy in a well-planned, well-organized and well-executed campaign against the FIMS Study! These activities included obtaining signatures on a petition against the study, statements read at the American Airlines annual stockholders meeting in Wilmington, Delaware, personal appearances on television and radio, and appearances at the New York City Department of Marine and Aviation.

The Manhattan Sloan House YMCA had a week long, summer program on *Problems of the Inner City*, in which they brought young men and women from all over the country to Manhattan to study urban community problems. They loved the FIMS controversy! The groups would go to Chelsea and be briefed by the Chelsea Against the STOLport Committee, and then come to American Airlines headquarters to hear my side of the story. I always reserved the dark mahogany-paneled American Airlines Corporate Conference Room for these meeting, served them soft drinks and cookies, and invited them to sit back in the big overstuffed leather chairs around the huge mahogany table and relax. I never said anything against anyone in Chelsea, and just explained what we were trying to do. These impressionable young people loved the attention!

The Chelsea Against the STOLport Committee held an organized demonstration in front of our headquarters in midtown Manhattan. It was a cold, windy, rainy day. The committee had prepared a "STOL

Chelsea Against the STOLport demonstration in front of American Airlines Headquarters

Dragon" of thin, plastic sheets, taped together, about 15 feet long, with a dragon's head, that they marched up and down in front of our building while the media (and our photographers) took pictures. There were a dozen or so people carrying signs: "Slay the STOL Dragon." When the wind tore the dragon's thin plastic body, I came to their rescue with Scotch tape and staples. My favorite sign read: "American Airlines; STOL Stooges of the FAA."

Our Public Relations department had two of our stewardesses, in full uniform, handing out hot coffee. One woman commanded loudly: "Don't drink their coffee! They are just trying to bribe us!" Jackie, always the voice of reason, told her to be quite: "They're doing no such thing! They are just being nice. I'm having a cup!" Which she did, gratefully. Of course the event was duly covered by the press and TV media, and thoroughly covered by the Chelsea press.

"Don't drink their coffee—they're just trying to bribe us!"

There had not been a big issue before FIMS for the elections in November, but there was now, and the candidates were not missing any opportunity to tell the voters what they wanted to hear. This is not a criticism of the candidates, but a simple statement of fact. Technical people must learn to understand and communicate with politicians who work and survive in a completely different environment than the sheltered position of engineers and scientists. STOL proponents were critical of elected city officials who had refused to build STOLports in response to neighborhood pressure groups.

This is unfair, unreasonable, and totally unworkable. It is the responsibility of elected officials to support the desires of their constituents. If the majority of the voices they hear say "we don't want a STOLport" and only the airlines, FAA, and manufacturers—none of whom vote in New York City—are saying "we want a STOLport," then the course of action is very clear; there should be no STOLport. The proper course of action for the proponents would be to determine if STOL is really better than the alternatives, and if so, to convince the people. This is where the U.S. supersonic transport failed, and this is where FIMS failed.

One of the concerns expressed by the Chelsea Against the STOLport Committee was a fear of a faceless FAA. They did not know the officials personally, and could not communicate effectively. To correct this problem, I arranged a meeting between the proper FAA and U.S Department of Transportation authorities and the Chelsea Against the STOLport Committee, which Jackie graciously hosted in her Chelsea brownstone home.

In addition to the FAA Associate Administrator for Plans and other STOL program office officials, I had invited representatives from the Federal, New York State, and New York City Environmental Protection Administrations. The Chelsea Against the STOLport Committee was there, as well as other key community leaders.

I explained our study program, and that this was only one small part of a much larger study that was investigating all aspects of the complex technical, operational, and economic questions that must be answered before any STOL decision could be made. Abruptly, one of the Chelsea residents stood up, interrupted my explanations, and demanded in a loud voice:

"Why don't you be quiet? We don't want you, we don't want your STOLport, and we don't want your study. Why don't you just go away and leave us alone?"

I had anticipated this reaction, if not so candidly or vehemently expressed, and was prepared with a carefully rehearsed invitation. I announced:

"The FIMS study *will* be completed. It is part of our obligation to our customers and to our stockholders to provide the best air transportation service possible. It is a technical feasibility study, and as a technical study emotional concerns would normally be inappropriate. However, since the feelings of the Chelsea community are so strong and will strongly influence any FIMS siting and future STOL operations, I believe your views should go everywhere the technical report goes. Therefore, if you will give me a statement I will include it, without editorial comment or modification, as an appendix to this report. Check your spelling because I won't change a thing."

This invitation was accepted, and I included the appendix in the final FIMS report. I only added a header and a footer identifying it as the opinion of the Chelsea Against the STOLport Committee.

Study Completed—FIMS Technically Feasible

The FIMS study was completed at the end of October, but I left it at the printers until after the November election in order to avoid exacerbating the controversy. The study found the FIMS technically feasible, with a single, 2000-foot STOL runway mounted above six World War II Liberty ship hulls arranged two by two. These ships were rusting away in an anchorage a few miles up the Hudson River. FIMS would have cost $14 million—far less than a land-based STOLport (and surprising close to my initial rough estimate of $12 million!). I sent copies of the report to the FAA, New York City officials, and to Jackie the following week.

Artist Conception of Floating Interim Manhattan STOLport (FIMS) in the Hudson River near Chelsea

With this published report finding FIMS technically feasible, the Chelsea reaction continued even stronger.

Working Relationship—American Airlines And Chelsea Against The STOLport Committee

Throughout the entire time my overriding position was one of open and absolute honesty with the Chelsea group. Furthermore, my contacts with the Chelsea Against the STOLport Committee and the Chelsea community were through Jackie. The one exception was a personal interview with Jim Buckley, Editor of the *Chelsea-Clinton News*.

The reason for this sole contact point was one of credibility. There must be no possible suspicion that I was trying to "divide and conquer," or trying to play one element against another. This would also help strengthen Jackie's position as leader and preserve the excellent communication channel that we had established.

I never tried to convince Jackie that I was right and that she was wrong regarding FIMS. I explained what I was trying to do, exchanged information, and that was all. There was no point in trying to change her mind. First of all, I doubted that was possible, and secondly, if I was successful then her committee would have simply rejected her and elected another leader.

The Chelsea Committee was perhaps more knowledgeable than most citizens' groups about the issues, because of the caliber of the community. They were still laymen, however, and did not understand many of the technical terms. Noise measurements were a case in point. Some measurements were made in db(A), others in PNdB (perceived noise level), and even EPNdB (equivalent perceived noise level). This was confusing. I tried to provide the committee with material to help them understand more of the technical aspects of the problems. Many times I would call Jackie and tell her of a study or research program that I thought she should know about. Some of these programs were not ready for public discussion, and I would caution her to keep this to herself, but that I thought she should know. On occasion she would call me for permission to report these things to her committee. We were on opposite sides, but I believed that Jackie was completely trustworthy, and would stand behind her word, and I knew she felt the same way about me.

One day I told her, in answering a question, that I would never lie to her, but sometimes I might not be able to answer her questions because we were opponents. She said that she knew that she could trust me. I said that I believed she did, but that perhaps some others in her committee with whom I had not dealt personally might not feel the same way. She said that they all knew they could trust me.

Jackie and I exchanged Christmas cards for ten years or so afterwards, and she accepted my invitation in 1975 to tell her story to my students at the University of Virginia, when I was a Visiting Associate Professor there. One of the biggest lessons I learned from this was to keep open the channels of communication with your adversaries, and maintain your complete candor and honesty—and sense of humor!

The Bucket of Noise

I learned an important lesson, from Jackie, about talking to lay people about technical subjects. I knew that the more complex a topic, the simpler it must be explained, but it really came home to me when Jackie and I addressed the American Society of Civil Engineers (ASCE) at Columbia University in Manhattan.

I had been invited to describe our FIMS study to the ASCE, but I accepted on one condition: that Jackie be invited to give her side of the story. She agreed, as a favor to me. I thought that it was important for the young engineering students to understand that even though their ideas and concepts might be technically feasible, economically viable, and operationally desirable, they had to also consider the sociopolitical implications. During my talk, I explained that: "The STOL aircraft noise level will be only 95 PNdB at 500 feet sideline The background street noise level is 95 PNdB. When you add 95 PNdB and 95 PNdB it's a logarithmic function, and the result is only 98 PNdB. Now, to *perceive* a doubling of sound level, the noise has to increase by 10 PNdB! So, obviously, the people will not be able to hear the STOL airplanes above the background street noise!"

Rob! You silver-tongued, glib, son-of-a-gun! You've done it again!

Then Jackie gave her story. She stood at the lectern and said, simply: "Well, I don't know anything about 'PNdB,' but it seems to me that it's kind of like water in a bucket. If there's water in the bucket, and you put in more water, the bucket gets fuller!"

I sat there with egg on my face, and resolved forever after to talk PNdB to technical groups, and "buckets of noise" to non-technical people. And for every briefing I gave after that, there was always someone in the audience who would request: "Rob, tell the 'bucket of noise' story!"

Public Hearing—FIMS Is Dead

The FIMS controversy came to a head on March 20th, 1971 with a public hearing, organized by Mr. Percy Sutton, Manhattan Borough President. It was held in the National Maritime Hall, International Association of Longshoremen, on Manhattan's lower Westside.

As the first scheduled speaker, I read a prepared statement describing the purpose of our STOL studies. I played a Dehavilland recording that compared the low noise levels of the STOL aircraft to the present generation of conventional jet airliners, and promised that American Airlines had no intention of "ruining a neighborhood." I summarized the many contributions that American Airlines made to social programs around the country to prove our good intentions. When we had flown the Breguet 941/MDC-188 along the Hudson River, made a u-turn at 500 feet above the water, and made simulated landing approaches along the Manhattan riverfront, including Chelsea, we received no complaints because the plane's engines were not heard above the background noise. Also, proponents of the 54th street heliport on the East River side of Manhattan had discussed their plan with a doctor whose office was right across the road from the proposed heliport. He agreed to let them fly some demonstration flights to show him that the noise would not disturb him. After he agreed to the demonstration, they casually mentioned that they had already been flying in and out of the site all morning, and he had not heard them. The heliport became operational.

Noise would obviously and demonstrably not be a problem.

The hearings continued the rest of the day, and many people, both pro and con, had their say.

The hearing itself was unnecessary as far as the outcome of the controversy was concerned, but it did provide closure for the cessation of hostilities. It did not affect the FIMS because Congressman Koch, at the start, read a letter from the U.S. Secretary of Transportation, John Volpe which read:

"It has already been determined there is no possibility of locating such a STOLport next to Chelsea in the foreseeable future. The strong opposition within the Chelsea community stopped further consideration."

This was an interesting experience. After the prepared statements, we speakers were questioned by a continually changing panel who made many short speeches with question marks at the end, which were, of course, unanswerable. In such instances, one should maintain his sense of humor, and keep his eye on the main issue. I had opened my prepared statement with a prepared ad lib that was appropriately quoted in the press:

"I feel like the turkey invited to Thanksgiving dinner."

I was featured on all three major New York City television networks' Six-O'Clock News that evening. Pretty heady stuff. I didn't know about the other interviews until I saw the news. They had interviewed someone in a long white coat that they introduced as a doctor, who announced that a Chelsea STOLport would essentially end life as they knew it!

Then, when they asked me about it, I stated simply and emphatically:

"Absolutely not! Tests show that the aircraft sound levels will not be heard above the normal street traffic!"

Short, concise, and to the point.

The following report from the May 1971 *Astronautics and Aeronautics* magazine, published by the American Institute of Aeronautics and Astronautics (AIAA), is provided in its entirety:

Manhattan STOLport: Cursed at Its Own Wake

It was an outpouring of irrationality, a shooting of fish in a barrel, this public hearing on a STOLport on the Hudson River shore in the Chelsea area of Manhattan. The Mayor's Council on the Environment sponsored it, but a member of the Chelsea Committee Against the STOLport chaired it. Arrayed on cither side of him on a raised platform sat a constantly changing panel of mostly elected officials who stuck needles into any witness who said a good word about a STOLport.

The chief stimulus for the hearing came from American Airline's feasibility study for a Floating Interim Manhattan STOLport (FIMS) in the Chelsea site. When American's Robin Ransone appeared to defend his FIMS, he said, "I feel like a turkey invited to Thanksgiving dinner." Following him in the witness chair came Jerome Kretchmer, Administrator of the City's Environmental Protection Administration, and Robert Rickels. Commissioner of Kretchmer's Department of Air Resources. While slamming the STOLport, they revealed the depths of their knowledgeability by deploring the threat of lead from gas-turbine fuel.

Those opponents who did not have empty heads had closed minds. After Ransone played a recording of the quiet STOL DHC-7 compared to jets, Ernie Patterson, head of the Chelsea-Elliott Tenants Association, said, "It didn't work on me . . . They want to put something else over on us ... I tell my kids, if it's about the STOLport, tune it out. You know what you want, and what you don't want."

In fact, the Chelsea Against the STOLport Committee could not seem to believe their own data on noise. One of the Against the STOLport Committee members had helped make a noise survey of Chelsea streets during the daytime hours. He found that the quietest spot read 64 dBA (about 77 PNdB), and the noisiest, 90 dBA (about 103 PNdB) when a truck passed. American had estimated that a STOL plane putting out 95 PNdB at 500 ft (Dehavilland now talks of 90 PNdB) would have a loudness of about 83 PNdB in the residential part of Chelsea nearest to the STOLport. Jerry Grey, new AIAA Director of Technical Services, who testified as a Chelsea resident, estimated 75-80 PNdB.

Thus, by the Committee's own readings, the quietest part of residential Chelsea may now be 2-dB noisier than the loudest STOL sound that would reach there. And the noisiest street is now noisier than it would be near the edge of the FIMS itself. The centerline of the FIMS runway would be 2300 ft from the nearest residence.

Blind Distrust

Unfortunately, American and the FAA, in the person of Jerold Chavkin, did nothing to break down the wall of blind distrust by trying to disclaim any designs on the Chelsea site. Chavkin, in his first public appearance since being appointed Acting Administrator of the V/STOL Special Projects Office, proclaimed his powerlessness. No steamroller, he. Chavkin made so many disclaimers that he sometimes seemed on the point of denying even the existence of the FAA. With a speaking delivery like Transportation Secretary Volpe's, he was a Sten gun of humility.

The proceedings turned into a postmortem when Congressman Edward I. Koch of lower Manhattan revealed from the platform a letter to him from Volpe saying, ". . . It has already been determined there is no possibility of locating such a STOLport next to Chelsea in the foreseeable future . . . The strong opposition within the Chelsea community stopped further consideration."

Questioned about the decision, Chavkin answered that the STOLport had been found "'unacceptable to the community."

A confused member of the interrogation panel kept asking what objective noise standard had been used to find it unacceptable. The two could not understand each other. For, of course, no objective standard had been used. By any objective standard, noise would not present a problem. Chavkin could only keep repeating that it was unacceptable because the people of Chelsea said it was.

Figures Lie . . .

Though the STOLport had become a dead issue for the present, witnesses continued to kick it. The honor of scaling the peak of absurdity fell to Percival Goodman, Professor of Urban Design at Columbia University, who compared the STOLport to Nazi cremation chambers and the My Lai massacre. Like most of the others, he offered no facts in evidence. "Figure lie, and liars figure," he said. And who needs facts and figures, when "on *prima facie* and *common-sense* grounds the STOLport is not a good idea."

To counter such examples of the new enlightenment. New York City Commissioner of Marine & Aviation Charles Leedham offered quiet logic. In a performance that towered over all others, he softly led through the chain of circumstances that has made the STOLport necessary for the city. The apprehension of dangers by Chelsea residents, not real dangers, has caused the STOLport's setback, he observed.

The city will now look to a site proposed by the Port of New York Authority in the New Jersey Meadows. When a working STOLport there has demonstrated its harmlessness, Leedham feels sure opposition will evaporate to returning to the Manhattan shore.
B.F.

Astronautics and Astronautics, May 1971

The Smithsonian Air and Space Museum Display—Confrontation of Technology With Society

At the time we finished our FIMS study, Mike Collins, the astronaut, and with whom I had flown one flight at Edwards, was planning the Smithsonian Air and Space Museum, in Washington, DC, and its exhibits. One of its exhibits was titled *Confrontation of Technology With Society*. Six recent cases where technology and society had clashed had been identified.

The exhibit sought to present the factors that either accepted or prevented the successful implementation of that technology. The case studies included: Personal Rapid Transit System—Morgantown, West Virginia; Mass Transit System—San Francisco Bay Area Rapid Transit (BART); High Speed Ground Transit System—United Aircraft Turbo Train; Automobile Transportation System—Automobile Simulations by CALSPAN; Sea Cargo Transportation System—Liquid Natural Gas (LNG) Tankers; and Air Transportation System—Manhattan STOLport (my FIMS study).

They chose FIMS because I had carefully documented everything that happened, had worked with the antagonists, and it represented all of the factors that seemed to be prevalent in such cases.

The firm of Booz-Allen, & Hamilton had the case study and exhibit design contract for the Smithsonian, and they interviewed Jackie Schwartzman and others of her Chelsea Against the STOLport Committee and me. I spent many hours telling them of my adventures, and providing documents and photos. They also interviewed Frank, and asked him, point blank: "Did Ransone *really* do all that he said he did?"

"Yeah, he sure did," was Frank's reply. And then he added: “Ransone is more than an engineer.”

They were amazed, and, in retrospect, I guess it was pretty incredible.

Years later, after completion of the Smithsonian Air and Space Museum, Paula and I saw the exhibit. It was in a Plexiglas case about eight feet tall and four feet square. There was a tree-like structure in the center, and ping pong balls were released from the top to roll down a number of chutes. There were holes in the chutes, labeled **Accepted**, through which the ping pong balls could eventually reach the **Success Box** at the bottom of the case—but there were also anti-technology factors that rotated about the case that could knock off the ping pong balls and plunge them, unsuccessfully to the **Defeated Box**. These factors were those that were common to all the documented case studies. It was fun to watch the visitors pulling for success, and watching hopefully as the ping pong balls fought their tortuous way through the obstacles.

Great fun.

The part of Booz-Allen, & Hamilton's report that pleased me the most, however, was an entry at the end of their report on FIMS. From their interview with Jackie, the authors reported: "...Ms. Schwartzman felt that Ransone was presenting them with good reliable information, and was a man of integrity…"

I thought that one sentence made all of my efforts with the Chelsea Against the STOLport committee worth while.

Thank you, Jackie.

Chelsea STOLport Case Study to the SAE

Fast forward to 1976 when I was a Visiting Associate Professor at the University of Virginia: As a member of the *Society of Automotive Engineers (SAE)*, I was invited to submit a paper for its National Transportation Meeting in New York City. The paper was to document my experiences with the Chelsea STOLport. I agreed on the condition that Jackie Schwartzman would be allowed to present her side of the issues. I still believed that it was crucial that engineers and city planners understand all of the socioeconomic and environmental factors for implementation—read "imposition"—of new technology on society and neighborhoods. Jackie agreed, as a special favor to me.

I had read a book about how to protest government or big business encroachment on personal space or activities. The book pointed out that most people join protests for purely personal reasons that have nothing to do with the issue at hand. These include a desire for publicity, job security, excitement of the argument, personal satisfaction, resentment at being pushed around, desire to go along with the crowd, personal political ambition, desire to follow orders, financial gain, and resistance to chance. And behind most such cases is fear—fear of the unknown, fear of change, fear of what it could lead to.

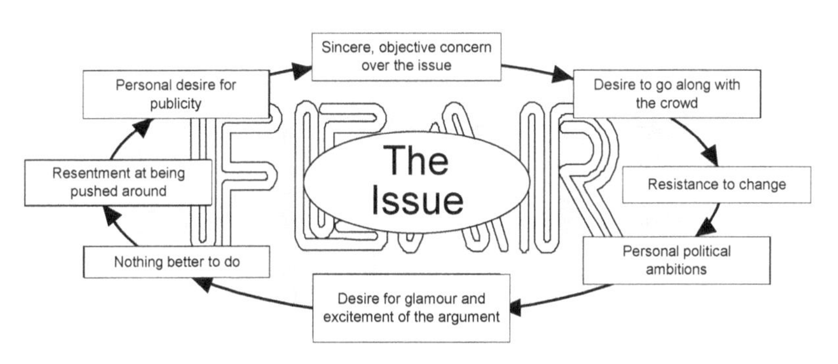

In addition to protest group's genuine concern for the issue, are very personal motivations, and behind all is fear

A few participants join the cause through a genuine concern for the issue, but if organizers want to muster a large, effective protest organization, they must cater to these personal motives as well. I included these in my paper because I had observed many—but not all—of these motivations during my work with the Chelsea Against the STOLport Committee.

I prepared a detailed history of the entire Chelsea STOLport confrontation, including what I was trying to do, how I interacted with the Chelsea group, and the outcome. I provided a list of *Lessons Learned* and advice for both sides on how to reach understanding and perhaps—but not always—agreement.

The three most important lessons learned for the proponents were: First, A complete and open explanation must be made at the earliest possible time, before rumors leak out with the wrong information. Second, the proponents must establish their credibility by speaking the truth, the whole truth, and nothing but the truth—silence or trick answers will destroy their credibility. And finally, responsible representatives of the affected citizens must be given a very real voice in the project planning.

But if the citizens group is afforded these courtesies, they must also accept some responsibilities: First, they must state their concerns openly and honestly. Second, if confusion arises because of conflicting statements heard out of context, they must determine the truth before reacting. Third, they must not react prematurely to comments made during early planning sessions. And finally, they must fairly represent the views of the citizens whom they represent, and keep them informed of what's taking place.

I concluded with a 13-point plan for undertaking any project that involved implementation of new technologies on society. The plan included inputs from local citizens groups. This was the only presentation of all that I had given that I read from a script. The reason was that I had 85 35mm color slides to present during my 20-minute presentation. I did not address these slides directly, but used the slides as counterpoints to my orals. For example, when I said that we completed the FIMS study and that it was proved technically feasible, I showed four slides showing the Chelsea newspaper headlines criticizing the study and the Chelsea Against the STOLport Committee demonstrating in front of American Airlines headquarters on Third Avenue.

Jackie gave her presentation, and it got a lot of attention, also, and, I hope, made the point I wanted to show. Jackie gave her presentation purely as a personal favor to me, and I greatly appreciated it.

The next year I was honored by having my paper, *Chelsea STOLport - The Airline View*, published in the prestigious *SAE Transactions For 1976* for its "high quality, lasting value, and contribution to the art." Only 10% of SAE papers presented each year are awarded that honor.

What fun!

The Propeller STOL Transport — the PST

Concurrently with the FIMS work, we also needed to know what STOL airplanes could be available in a reasonable length of time. Since the jet STOL would be too noisy, I wrote a brief specification for a Propeller STOL Transport (PST). I specifically asked for rough concepts because I didn't want the aircraft companies spending a lot of money until we narrowed the field down to those that made technical, operational, and economic sense. I released the Request for Information, it was duly reported in the media, and I received 13 proposals from the US, Canada, Japan, France, England, Spain, the Netherlands, and Germany. I evaluated these and narrowed the field to three possibilities: Grumman offered a C-130-type aircraft with blown flaps and control surfaces (similar to the C-130 Boundary Layer Control airplane that I had tested at Edwards). Canadair offered a four-engine, tilt-wing airplane, but in this version the wing tilt was only to 20-degrees, which greatly improved its STOL performance without overly complicating its flight control system with vertical flight requirements or a tail prop. And of course, Dehavilland offered its new four-engine turboprop Dash 7. Our marketing department loved the Grumman concept: "We can advertise that it has BLC!" they chortled.

Comparing the takeoff and landing distances was a bit complicated, each company had used calculations that showed its concept to its best advantage. Bob Baldwin and I at Edwards had published a paper defining STOL and VSTOL handling quality requirements. Our studies had shown that the higher the takeoff and landing lift coefficients, the more complicated the flight control system had to be for safe attitude control. Lift coefficients are basically non-dimensional measures of how much lift you can get out of a specific wing shape. The higher the lift coefficient, the slower the airspeeds can be for a given wing area. But you reach a point when the airflow over the control surfaces is not strong enough to control the airplane, and you have to go to controls that are independent of the free stream airflow over them, in fact, VTOL controls, even though you are flying STOL. I recalculated the takeoff and landing performance based upon these STOL concepts and criteria, and spotted the results on a chart.

I notified Grumman, Dehavilland, and Canadair that their concepts looked promising, and asked for rough orders of magnitude pricing and performance information. I notified the others of my selection, with my thanks. I gave each a copy of my assessment of their planes, and included a redacted copy of my takeoff and landing performance calculations. I showed all the airplane points, but identified only theirs. In due course, I got some rough pricing and performance information, which was helpful in assessing the state-of-the-art.

One company, however, was not happy with my decision. This was Fokker Aircraft in the Netherlands. Frank announced one day that their Vice President of Engineering and Vice President of Marketing were coming to see me.

Frank and I met them in the conference room, and we chatted a bit. They were very upset because I had kept Dehavilland's airplane and rejected theirs, which was the same size, and, they claimed, had identical STOL performance. Frank listened to all this, smiled sweetly at them, and told me "Well, Rob, it looks like you have another airplane to assess costs and performance!" He shook their hands, smiled at them, and left the room, leaving me alone with these too, very angry vice-presidents!

I explained how I had analyzed their airplane. They asked why I had kept Dehavilland's and rejected theirs, since the two airplanes were the same size.

"Well, for one thing, Dehavilland's airplane is brand new, utilizing the latest, light-weight structural techniques, and your Fokker F-27 is 20 years old, based on 1950 structural methods. It weighs 5,000 pounds more than Dehavilland's before you add any fuel or passengers! It can't measure up to Dehavilland's airplane, but if you want to waste your time and money in preparing cost and performance data, I will accept it, but I can tell you right now that it won't compare with Dehavilland's cost and performance characteristics."

They looked at each other, and slowly nodded. They thanked me, and left. I never heard from them again.

I went into Frank's office. "Thanks a lot for tossing me to the wolves!"

Frank just grinned. "I knew you could handle it," he said.

Frank loved stuff like this, and, frankly, so did I!

What fun!

The differences in takeoff and landing distances claimed by the airplane companies caused me great concern. If each company defined its STOL field length differently, there would be big problems for STOLport designers. As I explained it: “I don’t want to be flying an airplane with ‘STOL’ painted on my wings, over a field with ‘STOL’ painted on the runway, but I can’t land because my airplane and the runway are not compatible!”

So I wrote a paper that defined the *STOL Definition and Field Length Criteria,"* and presented it to the FAA for their consideration in rule making, and also presented it to the *AIAA 7th Annual Meeting and Technical Display* in Houston, Texas, in October 1970 (AIAA Paper No. 70-1240).

In this paper I clearly defined the terminal flight path, cross-wind requirements, glide slopes, and runway markings. I specified how the STOL pilot could determine whether he or she was operating within the safety margins, and what should be done during both takeoff and landing in the event the airplane strayed outside those safe boundary criteria. I also defined the STOL field length for takeoff and landing as 2000 feet. This last was based upon comprehensive work that Dehavilland had done in setting the field length for its proposed Dash-7 STOL transport.

Dehavilland had considered not only STOL aircraft performance and cost issues, but had also contacted several municipal authorities to determine what might be available. Five hundred foot-STOLports, no problem. Thousand-foot, 1,500foot, even 2,000-foot STOLports no problem. How about 2,500 feet? No dice. The municipal authorities told them they were nibbling them to death. So 2,000-foot STOLports were the criteria.

MacDac versus Boeing Airplane Company

The MacDonnell people that I worked with were all first rate, and extremely skilled marketers: I would get a telephone call, say, on Monday: "Hi Rob. Say, we'll be in New York on Wednesday and we have that information that you asked for. Will it be convenient for us to drop by?"

At 11:00 AM on Wednesday, Bob Brush, Vice President of Marketing and another person, usually an engineer, would arrive. Although we had overhead projectors in our conference room, they always came prepared with their own. This was before PC computers, LED projectors, and PowerPoint presentations. We would chat amiably a while, and then Bob would look at his watch and announce: "Say, it's almost lunch time. Why don't we grab a bite to eat and then come back to talk?"

That always sounded great! We at American Airlines had two lists of restaurants: one for when we were buying our own lunch, and another for when salesmen were buying. I'll leave it to you to guess which list was the more expensive! We would have a cocktail, lunch, and then get back to the office about 1:00 or 1:30.

There were always two people from MacDac. One would give the presentation and the other would watch us very intently. At a puzzled look or a raised eyebrow or a slight frown, he would stop the presentation. "Wait a minute, George. Rob, it looks like you might have a question."

I would explain my concern, and he would write this down very carefully. "Anything else? OK, George, continue."

They never left without some reason that I would want to hear from them again.

A week or so later, I would get a Monday morning telephone call: "Hi Rob. Say, we'll be in New York on Wednesday and we have that information that you asked for. Will it be convenient for us to drop by?"

I didn't always milk MacDac or Dehavilland for meals., but they wouldn't let me (AKA American Airlines) pay until I finally hit upon an argument that they couldn't refuse.

“Did you enjoy treating me to lunch?” I would ask.

“Yes, of course,” was the only answer.

“Then,” I would continue, “please don’t deny me that same pleasure.”

It always worked.

Boeing had a different, arrogant attitude entirely: Their whole approach was: “This is what we have decided that you need, and here is our airplane that will do it for you!”

On one of their visits, they described how their new STOL airplane design concept could meet our STOL field length requirement of 2000 feet. “It can takeoff and land in 2000 feet at a landing speed of 65 knots.”

“What kind of lateral-directional control system are you using?”

“Oh, just the usual drooped flaps and ailerons–simple, proven devices.”

“That won’t work, you realize.”

“Oh yes. Our analyses show that it will work fine.

It wouldn’t work at all. Bob Baldwin and I had researched the problem of low speed flight control at Edwards, based upon excellent work done in the 40x80 wind tunnel by our good friends at NASA Ames Research Center as well as their extensive flight tests of actual VSTOL research aircraft. To understand the problem and its solution, you need to understand the concept of “coefficients,” the aerodynamic analysis tool familiar to all aerodynamicists: A coefficient, as you know, is a non-dimensional function that can be determined from a small wind tunnel model and applied with excellent accuracy to full size aircraft. Lift coefficient, C_L, is an indication of how much lift you can get out of a specific wing shape. The total lift equals the lift coefficient times the wing area times the dynamic air pressure acting on the wing from the air flow over it. To fly a heavy airplane at 65 knots, the wing has to deliver much more lift (i.e., provide a higher lift coefficient) than at 85 knots, where the dynamic air pressure is much greater (OK, you aerodynamicists, I know it’s actually the lower static air pressure that really counts…). This higher lift coefficient is provided by such high lift devices as larger and double or triple-slotted wing trailing edge flaps, leading edge flaps (or slats), or propeller airflow over the wing. But since the dynamic air pressure is low, the rest of the airplane's control surfaces have to work harder, also, which requires different controls for slow speed flight than for higher speed flight.

NASA Ames had discovered from its wind tunnel tests that normal lift coefficients, say up to about 2.0, could be safely controlled with normal airplane controls of aileron, elevator, and rudder. Between lift coefficients of 2.0 and about 5.0, however, lateral-directional control augmentation was needed. This could be provided by spoilers (small slats or "fences" that could be raised along the middle of the wing to spoil the lift and drop that wing without yawing the aircraft). Drooped ailerons would not work because, although they could provide the needed lift, when they were activated they caused serious adverse yaw due to the changes in asymmetric drag from the aileron deflections.

NASA Ames also discovered that above lift coefficients of 5.0, full VTOL flight controls were required about all three aircraft axes—essentially the controls had to be independent of free stream air pressure, provided by forces generated by the aircraft propulsion system.

So I just smiled at the Boeing people, smug in my knowledge of things that these arrogant people had not yet discovered for themselves. I remembered what my old A&M aircraft structures professor used to say: "It's not what I *don't* know that bothers me—it's what I *do* know that *ain't so*!"

Six months later, these same people came back to up date me on their STOL aircraft design.

"What is your landing speed for a 2000-foot landing?"

"75 knots," they replied. Outside of the fact that 75 knots would require uncomfortable braking for the passengers, I let that go."

"And what type of flight control system do you have?"

"We have full-span trailing edge, triple slotted flaps, leading edge slats, and spoilers along the top of the wing," they answered quietly.

"You've been in the wind tunnel, haven't you?"

"Yes," they admitted. They would have answered sheepishly, but Boeing Airplane Company was never, ever, sheepish!

What fun!

The VSTOL Space Ship

I got lot's of telephone calls and STOL material from complete strangers who read of my work in the technical journals and news media. Some of it was useful, but most wasn't, and one, in particular, was really wild! I got the call one afternoon.

"Hello. This is Rob Ransone. What can I do for you?"

"Are you the Rob Ransone in charge of American Airline's VSTOL programs?"

"Yes."

"I hear you have requested information on propeller-powered STOL airplanes."

"Yes, that's right, because the jet engines are too noisy."

"Well, I have a STOL airplane that can fly as fast as a jet but is as quiet as a propeller."

"Oh?"

"It's not only STOL, it is also VTOL.

"Oh?"

"It's a modular airplane, with sections of fuselage that can be bolted into place or removed depending upon how many passengers you have. You never fly around with empty seats!"

"Wow."

"It's not only as fast as jet STOL planes, but it is supersonic and hypersonic!"

"Incredible!"

"In fact, it can fly in outer space!"

"Hm..... What kind of engine does it have?"

"A high speed diesel engine!"

"Just a minute."

I put down the phone, walked out of my office, leaned against the wall and laughed until my sides hurt. Regaining my composure, I returned to my office and picked up the phone. I didn't ask him what the high speed diesel engine was attached to that gave such incredible performance because I knew that I would have to put down the phone again.

"Have you actually flown this airplane?" Of course, I already knew the answer.

"Not yet."

"So it's still in the experimental stage?"

"Yes."

"Well, that wouldn't be developed enough for us, I suggest you contact NASA."

"OK, thank you!"

It's days like that that make going to work worthwhile!

Two years later, when I was working at NASA, I told that story and got a different reaction:

"So you 're the SOB who turned that weirdo onto us! We told him to call the airlines!"

Airline View Of STOL System Requirements

One day one of the fellows from Corporate Planning was in my office complaining because his boss wasn't giving him anything to do. I had always looked ahead to identify the next activity needed for my job, so I really lit into him:

"He should probably fire you," I challenged him.

"Wha.. .what do you mean?" He stammered.

"Look, you should be thinking ahead about what the company needs that you can provide. You should have so many ideas cooking on the back burners that whenever you finish a job you should dash into your boss's office and demand more help in doing your job, and explain why that job is important to American Airlines! Your boss would be grateful, and in no time you would be running things!"

He thought about this for a bit, and then left my office. He wasn't too happy with me, but I hope it gave him some incentive that his boss was too preoccupied to provide. I understand that a few years later American Airlines abolished the Corporate Planning office as an economy measure.

Along this line, as I was finishing the FIMS study, I thought ahead to what was next. I figured that, with all my work and the information that I had accumulated, that I knew more about what was needed for a successful STOL airline service than anyone, so I sauntered into Frank's office with a suggestion that I conduct a thorough study of what it would take to bring short-haul STOL airline service to fruition. This would be a comprehensive systems engineering study years before the term "systems engineering" became a buzz word.

Frank liked the idea, but American Airlines, like the other domestic airlines in 1971, was having some financial difficulties. They couldn’t afford it. So I prepared a presentation and flew down to Washington to talk with my friends in the Department of Transportation. To give you an idea of the support I had from my friends at American, Fran Nadolna, Frank's secretary, lit a candle for me at her Catholic Church, and Ellie Andrews, our receptionist, prayed for me on bended knee! What a support group!

Ellie was special: Not only very, very attractive, but she displayed her long legs in the very short micro-mini skirts popular during the late 1960s. A young widow, she had lost her husband, a New York City fireman, when a burning building collapsed on him. She was also very attracted to me, and, I'll admit, that I found her also very attractive, and we flirted frequently. One day she asked me: "Rob, do you run around?" It was not an invitation, because she did not mess around with married men, but just a simple question.

"No," I answered immediately. "Because there is nothing that I can think of that another woman could offer that would be worth risking what I have at home."

She gave me the biggest smile I ever saw, and confided: "You just went right to the top of my list!"

I flew to DC and met with Dan Green and Dick Halpern at DOT. Dan was a Program Office Director and Dick was an Under Secretary. I gave my presentation of what I wanted to do, and explained that, since American couldn't afford me anymore, that if they didn't support this study, that I would be out of a job, and American would terminate its STOL studies. This last bit was the clincher: later, Dan told me that they didn't want American to stop working on STOL, and they gave me a contract to conduct the study.

I solicited the specialty inputs of others at American and published the study in February, 1972. It defined 23 marketing, economic, STOLport siting, technical requirements, and environmental conclusions, and 34 specific recommendations.

DOT was pleased.

Frank was pleased.

General Electric was *not* pleased. Here's why:

GE Aircraft Engines was one of many companies working on an FAA specification for STOLport sites. I had written an AIAA paper on *STOL Definition and Field Length Criteria* that defined, specifically and in measurable terms, exactly what constituted STOL aircraft takeoff and landing performance, and the 2000-foot field length criteria needed for safe operations. It also included maximum noise levels that would likely be accepted by neighboring commercial and residential communities. The FAA task included representatives from the airlines, aircraft and engine manufacturers, international aviation organizations, and airport managers as well as the FAA and DOT. It was an august group. I was American's representative. Everything was going fine until we got to the sideline noise level specification.

GE wanted to define it as 106 PNdB sideline noise at 500 feet because that was the best they could do with jet engines. I argued for 95 PNdB because that is what had been accepted by heliport neighbors for the Sikorsky S-61 transport helicopters. I argued that 106 PNdB *might* be OK, but we didn't know that, and we would not spend several hundred million dollars on STOL airplanes and then find out we couldn’t use them. GE argued that STOL engines would be noisier than conventional jet engines because they had to be more powerful. Then I asked how come Cadillac engines are quieter than Harley engines? GE just glared at me!

I was losing the argument, until I announced that if they insisted on 106 PNdB, then I would withdraw from the committee and publish my own specification establishing 95 PNdB and denouncing their 106 PNdB specification! And at that time I had more national and international VSTOL credibility than General Electric.

This "2 x 4" treatment finally got their attention, and I prevailed.

I was an arrogant SOB.

GE was miffed.

...But Frank was pleased!

Advanced Medium VSTOL Transport

About 1970 the U.S. Air Force released a Request for Information (RFI) for a medium sized, turbo-prop VSTOL transport to augment the C-130s. Boeing Vertol, in Philadelphia, was responding to the RFI and invited me to come see what they had and how it might meet our commercial requirements. Marcy Fannon, from American Airlines, and I caught the *Metroliner* from New York City's Penn Station to Philadelphia. This was a very comfortable, high speed (100-110 mph) luxury train run by Penn Central until taken over by the National Railroad Passenger Corporation (Amtrak) in 1971.

Vertol had a reasonable design, but its presentation had some STOL data mixed up with VTOL data, especially with regards to noise, which I recognized. The presentation showed how propeller RPM could be reduced to reduce the sideline noise levels.

I asked: "How much can you reduce the propeller RPM during a vertical landing without incountering propeller blade stall?"

There was a very long silence, and then someone in back cleared his throat and admitted. "Well.....none......"

I knew this because we had had a very hard landing with the XC-142A at Edwards when the pilots forgot to increase the propeller RPM from 95%, used in cruise, to 100% for the vertical landing.

Broke the aircraft in half!

I wrote an assessment of Vertol's concept and sent it to them. Basically it said that the concept was very good, but noise would be a show-stopper for vertical operations in noise sensitive areas, although the presence of cross-shafting between the engines would cause no more additonal maintenance expense that an extra propeller. This we had learned from our tests of the Breguet 941/MDC-188.

Vertol agreed with my assessment, and added that they had suspected this, but that it was a bit painful to see it all written down.

What fun!

A Romp With the Big Guys

My work at American Airlines was very high profile, as you have probably noticed. I was frequently quoted in the local press and in international Aviation's weekly "bible" *Aviation Week and Space Technology*. I gave many briefings and papers to aviation groups, city planners, technical societies, NASA, FAA, and DOT. Several briefings stand out, in particular.

One of the first briefings was to the U.S. Air Force Scientific Advisory Board. This was an august committee of high level Air Force and industry executives that advised the U.S. Air Force on policy and procurement decisions. I prepared view-graphs, my presentation was well received, and I answered all of their questions.

Another important briefing was to Secor Brown, chairman of the Civil Aeronautics Board (CAB). This was the powerful government agency that regulated the airlines, determining which airlines flew where and what they could charge, before deregulation changed all that and resulted in the dissolution of the CAB. With the possible exception of the President of the United States, Secor was the only official in Washington that George Warde, American's president at that time, wanted to accompany me for the meeting. Although I had laryngitis, the briefing went well. Secor had lived in Japan for several years and had many Japanese mementos in his office. Since I had lived in Japan for 2 ½ years shortly after the war with my mother and sister, while my dad, an Army officer, was stationed there with the occupation, I could tell him "hello" in Japanese, and ask him how he was, and discuss some of his treasures. He was relaxed and hospitable, very interested in what I was doing at American, and I thoroughly enjoyed our meeting.

George was pleased.

I also briefed Neal Armstrong. I had known (of) him when I was at Edwards because of his rocket research plane flight tests at NASA, Edwards. After his moon flight he was assigned to NASA Headquarters in Washington, D.C. He was interested in our STOL program, so I flew down to Washington and briefed him. He was very casual and personable, and asked all the right questions.

Paula was disappointed with me when I got home because I had not asked him about his moon flight. I explained that we had talked only about our STOL programs, and I had felt awkward about asking him: "Oh, by the way, how did you find the moon?"

My next really interesting briefing was a few months later. Neal had left NASA, and had been replaced by Dr. Wernher von Braun, the German rocket scientist who had developed the V-l "Buzz-bomb" and the V-2 missile at Peenemunde. Dr. von Braun also wanted to know about our STOL programs, so I few to Washington and briefed him.

Dr. von Braun was also personable and interesting to talk to. I had mentioned the oft repeated cliché that we would need the STOL aircraft before STOLports would be built, but we would need the places to land before manufacturers would build the STOL airplanes. "It's the old 'chicken and egg' syndrome." He nodded his understanding. "By the way, " I added, "I know the answer to which came first, the chicken or the egg."

"Oh?" He was interested.

"Yes. The *rooster* came first—*he* laid the hen!"

Well that broke him up! He had a great sense of humor and enjoyed that little joke.

What fun!

One day Bob Houghton, President of Lockheed Georgia aircraft company visited Frank and me to find out about our STOL work. After a few discussions, he told Frank: "I've been trying to decide who is really driving the whole airline STOL effort."

Frank pointed to me, and said "Ransone is."

Bob nodded, and replied "That's what I've decided, too."

But there was one thing that Bob was more excited about than STOL. The day before he had visited Charlie Blair, president of Antilles Air Boats, a short-haul Caribbean airline that served the American Virgin Islands with amphibious Grumman Widgen airplanes. He had called at Charlie's New York apartment, and had been greeted by Charlie's wife, who was most cordial, and served him a cocktail and chatted with him until Charlie returned. Bob couldn't wait to telephone his wife in Georgia to gloat over who had served him a cocktail that afternoon! Did I mention that Charlie Blair's wife was the gorgeous, red-haired Hollywood actress Maureen O'Hara? Little did Charlie Blair know, but it was *I* who saved Maureen from those dastardly pirates from my vantage point in the front row of the theater years ago!

Dick Halpern, DOT Under Secretary for Aviation, conducted hearings on short-haul STOL service, and invited Scott Crossfield from Eastern Airlines and me from American Airlines to brief him and his staff on our work. Scott was more famous, so he went first. That was OK, except that he talked on and on and on and on..... In fact, he talked so long that Dick ran out of time and I didn't get to speak. I told Dick that I wanted another chance, because we at American had discovered some different results than Scott had found at Eastern. Dick was receptive, an I got another chance a few weeks later. Scott reported that Dick was very impressed with both of our presentations, and thought both of us had a lot of common sense.

A few months later at an American Institute of Aeronautics and Astronautics (AIAA) meeting in DC, DOT presented another approach to high speed short haul transportation. This was for a high speed rail system, and one of the main panel members was the DOT Under Secretary for Rail Transportation. He explained that the main problem with very high speed rail transportation was the very high aerodynamic drag of a half-mile long train traveling along the ground at very high speed—like 500 mph. Yes, this is very true. In order to solve this problem, his consultants had conceived a novel approach: The train would travel through a long tube, or tunnel, from which the air had been evacuated. Since the train would be traveling in a vacuum, there would be no air drag. This is also true. The route would be from New York City to Dallas, Texas, because that route, running through mountains instead of over them, would not have to travel up hill. There would be air locks at the stations in order to preserve the vacuum. Maybe. Since this would obviously be a very expensive undertaking, it would require a US Government subsidy of three billion dollars every year! Uh oh!

Man, in the "nut" category, this concept was right up there with the "accumulator airplane" and the "high-speed diesel-powered space ship!"

Dick was obviously of the same opinion, and asked the large audience if they had any questions. I was sitting in the front row, and he looked right at me, pleading... "Doesn't anyone want to say anything about STOL?"

I was never shy about asking embarrassing questions, especially to pompous Government officials. I raised my hand, and Dick introduced me, and announced who I was.

I stood up and tossed my bombshell: "Well, as a strong STOL proponent, I am greatly encouraged by what this gentlemen has said about high speed rail, but as an American taxpayer, I am greatly disturbed! He has stated that this high speed rail system will require an annual Government subsidy that is greater than the entire development costs of the U.S. supersonic transport that was canceled because it was too expensive. As a taxpayer, I would like to know what you are doing to keep my taxes down!"

The entire audience applauded.

The expression on the face of the DOT Under Secretary for Rail Transportation indicated that he really wished he was somewhere else, or that I would suddenly fall through the floor! Finally he stammered something about how, of course, they were always concerned about taxpayers' dollars....

Dick smiled, satisfied that I had rained on the railroad guy's bullshit parade.

Frank was pleased.

What fun.

American Airlines—Cleared For Departure

After all this work, what happened to STOL and FIMS? The American Airlines STOL studies, including my landmark report of *Airline View of STOL System Requirements* under U.S. Department of Transportation contract, were completed in February, 1972. This included studies of aircraft flight evaluation, FIMS technical feasibility, operational aircraft proposals from industry, very detailed market and economic evaluations, community acceptance concerns, and overall American Airlines corporate policy. The FIMS concept itself, exclusive of siting, would have been feasible technically, operationally, and would not be heard over the city's background noise levels. It would have been marginally feasible economically but only with significant U.S. Government or trunk airline financing.

A few years later, during the 1970s, Canadair provided short-haul commercial flights between Ottowa's Rockcliffe Airport, only 4 nautical miles from Downtown Ottawa, and the Victoria STOLport near downtown Montreal. The Montreal STOLport was located on a garbage dump island near downtown Montreal. The AirTransit objective was to demonstrate Twin Otter STOL aircraft in downtown areas and avoid longer drives to the Ottawa and Montreal airports. The demonstration was well received and utilized, but was terminated after two years in accordance with Canadair's original agreement with the cities for only a temporary demonstration.

Dehavilland Twin Otter in Air Canada* AirTransit *livery

The Chelsea site for FIMS was not acceptable because of the community reaction, and also because of questionable ease of access from the middle and upper East side of Manhattan. The *Buffalo* STOL service demonstration was not economically feasible due to the high costs of the aircraft modifications and certification, and their planned short operational life.

The operational Propeller STOL Transport (PST) would have been technically, operationally and economically feasible between a Manhattan STOLport and a Washington, DC STOLport. American Airlines decided not to implement STOL at that time, however, for other reasons related to its corporate objectives, overall airline economics in 1972, and long range plans.

The only visible result from all this effort was that Dehavilland produced the quiet, 48-passenger turboprop powered Dash-7 STOL transport aircraft that instituted short haul service in many locations across the country. I was invited to the formal roll-out ceremony at Dehavilland Aircraft in Toronto, Canada. Although not the Father of the Dash-7, at least I considered myself the "milkman."

My conclusions about short-haul STOL service at American had shown that it would not be economically viable at that time, so I had essentially worked myself out of a job. I had given a briefing of my studies to a NASA Short-Haul Conference in New Hampshire, and captured the attention of George Cherry, Director of the Aeronautical Operating Systems Office, in the Office of Aeronautics and Space Technology, at NASA Headquarters in Washington, D.C. George had been an electronics engineer at MIT's Draper Labs in Boston, and had written the computer algorithms that enabled Neal Armstrong to land safely on the moon in 1969. When Neal went to work at NASA Headquarters, he set up the Office of Aeronautical Operating Systems (essentially airline research studies) specifically for George. Now, Neal had left, and Roy Jackson had assumed the title of Assistant Secretary, of the Office of Aeronautics and Space Technology, and had promoted George to be his Deputy.

Now George needed a replacement, so he offered me his old job. This would be an Excepted Service position in the Civil Service, and took only two weeks for my appointment. George promised me big things. The perks included a reserved parking place in the NASA Headquarters parking garage, a huge office on the third floor overlooking the Smithsonian Museum and the Washington Mall, and soon a huge salary increase. It also seemed like a good time to leave New York. I had gotten tired of the traffic and congestion—especially after I smashed my briefcase across the side window of a car that tried to run me over in a crosswalk! The driver had looked as surprised as I did, and we both hurried on.

George Warde, now president of American, told me that if I wanted to come back he would make a job for me.

American Airlines going away party

So, in May, 1972, Paula, Key, Cheryl and I took a last free trip to the Caribbean, we said goodbye to our good friends at American, listed our beautiful new house in the northern New Jersey woods with a realtor, and I drove to Washington, DC.

I should have taken George Warde's offer...but that's another story...

Table of Contents